AF696408

ECONOMICS OF TECHNOLOGICAL CHANGE

ECONOMICS OF TECHNOLOGICAL CHANGE

THOMAS L. BEANE
EDITOR

Nova Science Publishers, Inc.
New York

Copyright © 2007 by Nova Science Publishers, Inc.

All rights reserved. No part of this book may be reproduced, stored in a retrieval system or transmitted in any form or by any means: electronic, electrostatic, magnetic, tape, mechanical photocopying, recording or otherwise without the written permission of the Publisher.

For permission to use material from this book please contact us:
Telephone 631-231-7269; Fax 631-231-8175
Web Site: http://www.novapublishers.com

NOTICE TO THE READER

The Publisher has taken reasonable care in the preparation of this book, but makes no expressed or implied warranty of any kind and assumes no responsibility for any errors or omissions. No liability is assumed for incidental or consequential damages in connection with or arising out of information contained in this book. The Publisher shall not be liable for any special, consequential, or exemplary damages resulting, in whole or in part, from the readers' use of, or reliance upon, this material.

Independent verification should be sought for any data, advice or recommendations contained in this book. In addition, no responsibility is assumed by the publisher for any injury and/or damage to persons or property arising from any methods, products, instructions, ideas or otherwise contained in this publication.

This publication is designed to provide accurate and authoritative information with regard to the subject matter cover herein. It is sold with the clear understanding that the Publisher is not engaged in rendering legal or any other professional services. If legal, medical or any other expert assistance is required, the services of a competent person should be sought. FROM A DECLARATION OF PARTICIPANTS JOINTLY ADOPTED BY A COMMITTEE OF THE AMERICAN BAR ASSOCIATION AND A COMMITTEE OF PUBLISHERS.

Library of Congress Cataloging-in-Publication Data
Available upon request

ISBN 10: 1-60021-237-9
ISBN 13: 978-1-60021-237-6

Published by Nova Science Publishers, Inc. ✢New York

CONTENTS

PREFACE

This book brings together a collaboration of technological advancements and how they have influenced the world's economy. These include production efficiency, industrial innovation, regional developments and issues economic growth.

Two themes dominate the contours of world development—progress and divergence. Rich and poor countries are diverging into two "clubs"—the progressive and the stagnant. Endogenous growth and multiple-equilibria models seek explanations through factor accumulation, externalities and non-convexities. Chapter 1 focuses instead on effects of learning on total factor productivity, and we show that growth divergence reflects a failure to learn. We use psychologists' Yerke-Dodson Law to upgrade the Gerschenkron-Nelson-Phelps linear learning model: economic backwardness (backlogged technologies) helps learning only up to a point. The upgraded model explains both growth divergence and club-convergence. Policies regarding education and appropriate technology are derived.

Since Schumpeter (1942), the need to provide the successful innovator with some form of patent protection in order to stimulate investment in R&D has been widely recognized. As patent protection helps innovators to benefit from their research efforts, it creates or increases the incentives to perform research activity. It also promotes technological progress and, thus, economic growth.

The analysis of optimal patent lifetime is founded on two branches of economic theory that in recent years have proceeded, for the most part, separately.

In the IO literature, the analysis is built on the hypothesis that a society must balance the gains accruing from rapid technological progress against the welfare losses associated with the presence of temporary monopolies in the use of a new technology (Nordhaus, 1969; Loury, 1979). However, this literature is mainly confined to study the incentives and the distortions induced by a patent system within a static, partial equilibrium framework. Judd (1985) represents the first attempt at studying the optimal patent lifetime through the maximization of a social welfare function in a dynamic, general equilibrium set-up. He builds an exogenous growth model where innovation, though endogenous, is not sustainable when there is no exogenous increase in the labor force. Still, his analysis remains important because it shows that patent lifetime can have important effects on the equilibrium innovation and growth rates.

Within the endogenous growth literature, innovation-based economic growth has become an important field of research due to the pioneeristic works of Romer (1990), Grossman and Helpman (1991) and Aghion and Howitt (1992). In these models, though, no attention is paid to the patent system as a *government policy tool* (it is assumed that a successful innovator always accrues an infinite patent protection against future innovators). More recently, Michel and Nyssen (1998) have analyzed the role of patent lifetime on growth in a similar framework. Nonetheless, their analysis remains limited as they consider a deterministic innovative activity. Loury (1979) was among the first to highlight the limits of modeling innovative activity as a deterministic process, instead of a stochastic one. On the other side, Aghion and Howitt (1992) and Grossman and Helpman (1991, Chapter 4) assume that innovative activity is stochastic but continue to maintain the assumption that patent life is infinite.

Our motivation in Chapter 2 is to fill the literature gaps mentioned above. Using an endogenous technological change model with vertical differentiation and stochastic R&D activity, we address the following question: what is the patent lifetime a government would set in order to maximize aggregate economic growth? Within this framework we also study how this patent length would change in response to changes in the degree of competition in R&D, the productivity parameter of the innovation *production function* and the level of innovator's monopoly profits.

Chapter 3 empirically examines the relationship between knowledge capital and performance heterogeneity at the firm level. The model is based on a knowledge production function comprising of four interdependent equations linking innovativeness to innovation input, innovation output and productivity. The empirical part is based on Korean firm level innovation data. The model is estimated using advanced econometric methods. We investigate whether innovation is a significant and contributing determinant of performance heterogeneity among firms. In examining the relationship between innovation and productivity we correct for selectivity and simultaneity biases. The results show that there is a two-way causal relationship between knowledge capital and labor productivity. Firm-specific effects positively contribute to innovation output but they are negatively related to productivity. Industry heterogeneity does not affect innovation output or productivity.

Recent years have seen growing interest in research and development (R&D) in Australia. Unfortunately, little is know about the contributions of R&D to Australian economic performance. Even less is known about the contributions of university R&D. Chapter 4 examines the impact of both university and non-university R&D on output levels in Australia, using data for Australian states, for the period 1979 to 2003. Both Cobb-Douglas and Translog versions of the production function are estimated. The results indicate that while both types of R&D make important contributions to the generation of output, not all university R&D has a positive effect. Pure basic and applied university R&D both have had a positive effect on regional output. However, this is in sharp contrast to the impact of strategic basic and experimental R&D, which are shown to have a negative effect on output.

In the 1980s, the United States established patent protection for biotechnology. Following this, patent applications for agricultural inventions have grown significantly. Nonprofits produced many significant agricultural innovations in the past. Now for profits are producing an increasing share of U.S. agricultural innovations. Two measures of innovation are considered: the number of new agbiotech patents and plant variety protection (PVP) certificates. Chapter 5 explores the similarities and differences in the roles nonprofits and for-profits have played in agricultural innovation in the past 40 years. The data reveals that for-profit organizations combined own greater numbers of patents and PVP certificates than nonprofits combined. However, the average nonprofit organization owns more patents than the average for-profit organization. The average nonprofit is also likely to hold more patents in DNA technologies than the average for-profit. Finally, nonprofits overall have created more plant varieties than for-profits in particular crop types such as oats, potato, peanut, and rice.

Indian Railways (henceforth referred to as IR) is a Department of the Government of India (henceforth referred to as GOI). The control mechanism between GOI and IR leaves IR in a confused situation as far as specification of its objectives is concerned; it is expected to be financially viable while at the same time fulfilling certain social obligations, which are not supported by GOI. Also during the last few years, financial support from GOI has been on the decline thereby adversely affecting the financial position of IR and leaving it to fend for itself by means of market borrowings. In such an environment, it is imperative that IR operates efficiently, i.e. use the inputs at its disposal to get maximum output from the same. IR initially had nine zones many of which were further split to form sixteen zones. GOI and IR justified this division of zones in terms of increase in efficiency due to smallness of size. This paper attempts to find an empirical justification for bifurcation of the zones. In Chapter 6, the Data Envelopment Analysis (DEA) technique has been used to study the productive efficiency of IR. This is essentially a non-parametric technique, which does not require restrictive assumptions of cost minimization, which was essentially a limitation even with flexible form cost functions such as the transcendental logarithmic (translog) cost specification. Specifically, since panel data (for nine zones and time period 1989-90 to 1999-2000) was available, the Malmquist DEA technique was used to decompose Total Factor Productivity (TFP) into two components; one part due to efficiency change, which indicates to what extent firms are getting closer to the frontier and technical change, which captures shifts in the frontier itself. In the context of IR, technological change would mean changes in quality of track, wagon and locomotives with respect to production efficiency. This exercise indicated that zones of IR are not making optimal use of the inputs at their disposal. Also larger zones of IR, in terms of output (passenger kilometres and ton kilometres), seemed to be relatively efficient as compared to smaller zones. This seriously challenges GOI's decision to split IR into smaller zones, which seems to be devoid of any economic rationale. Another interesting fact emerging from the study was that while zones of IR made the best possible use of whatever inputs were made available to them, IR as an organization has done precious little in embracing cost minimizing technology.

Chapter 7 analyses the relationship between technological collaboration and collusion. Firms can collude or defect on the output market, but with the novelty that there is a possibility of developing a new technology jointly. The development of the new technology is conditional on prior collusion by firms. The paper derives three main results. First, it is found that technological collaboration facilitates product market collusion, but only in the short-run, by creating an equilibrium, where firms collude initially in order to develop the new technology, and then defect afterwards. Nonetheless, there is less defection compared with the standard trigger strategy model. Second, it is shown how the equilibrium depends on the discount factor; in particular, delayed defection is sustained for intermediate ranges of the discount factor. Finally, the comparative static analysis shows how the equilibrium is affected by changes in the environment. An increase in the scope of the market, an increase in the number of firms, and a lower quality technology all contribute to making collusion more difficult. Whereas permanent collusion is affected only by changes in the number of firms, delayed defection is affected also by the scope of the market and the quality of the new technology. The results are shown to extend to Bertrand competition, although in the latter case there is less delayed defection than under Cournot competition.

In: Economics of Technological Change
Editor: Thomas L. Beane, pp. 1-22
ISBN: 978-1-60021-237-6
© 2007 Nova Science Publishers, Inc.

Chapter 1

LEARNING AND GROWTH DIVERGENCE

H.M. Leung*

School of Economics and Social Sciences,
Singapore Management University,
90 Stamford Road, Singapore 178903

ABSTRACT

Two themes dominate the contours of world development—progress and divergence. Rich and poor countries are diverging into two "clubs"—the progressive and the stagnant. Endogenous growth and multiple-equilibria models seek explanations through factor accumulation, externalities and non-convexities. We focus instead on effects of learning on total factor productivity, and we show that growth divergence reflects a failure to learn. We use psychologists' Yerke-Dodson Law to upgrade the Gerschenkron-Nelson-Phelps linear learning model: economic backwardness (backlogged technologies) helps learning only up to a point. The upgraded model explains both growth divergence and club-convergence. Policies regarding education and appropriate technology are derived.

JEL Classification Numbers: I20; O10; O30; O40

Keywords: Growth-divergence; Convergence clubs; Twin peaks; Relative backwardness; Yerkes-Dodson Law; Learning; Appropriate technology

"Learning is like rowing against a current—as soon as you stop, you drift backwards."
– Philosopher Lao-Tse (604 B.C.)

I. INTRODUCTION

Two themes have dominated the contours of world development since the past millennium. The first one is progress, and the other is divergence. Maddison (2001), in his exemplary

* E-mail address: hmleung@smu.edu.sg, Phone: (65) 6828 0717

portrayal of the world economy, recounts that the world's per capita income increased 13-fold in the past millennium, but that the per capita income ratio between the rich and the poor countries diverged dramatically – from about 1:1 in 1000, to 7:1 in 1998. This twin-momentum of growth and divergence is relatively modern. In terms of growth, Maddison describes the advance in world per capita income from 1000 to 1820 as a "slow crawl" (p.17), followed by spurts of great economic conquest (as well as technological and colonial conquest). Such conquests were led first by Britain, then by Western Europe, and in the past century by the United States. In terms of divergence, Lant Pritchett (1997) devises a "backcasting" imputation method, which produces even more distressing results. He argues that the per capita income ratio between the richest country (USA) and the poorest country (Chad and Ethiopia) was 8.7:1 in 1870, but 45:1 in 1990 (1997; p.10).

No matter what intellectual school of thought one belongs to, the observed trend of world income divergence is both compelling and troubling. The most disturbing part is that the income of Sub-Saharan African countries, which make up three quarters of Africa's population, fell by a quarter from 1980 to 1998 (Maddison, p.163). The gap is widening, and the plight of the poorest shows no signs of relenting.

Still, the cloud has a silver lining. This comes not from Africa, but from what Maddison calls "Resurgent Asia" – China, Hong Kong, Malaysia, Singapore, South Korea, Taiwan and Thailand. "Resurgent Asia" was preceded by Japan; Japan's industrialization began with the Meiji Restoration in 1867, and she became the first Asian industrial miracle after WWII. From 1973 to 1999, the average per capita income of the 7 Resurgent Asian economies, excluding Japan, grew at 5.5% per annum, which is 2.6 times faster than that of the United States, and 3.3 times faster than that of Western Europe. This catch-up is swift and impressive. In 1960, Singapore's per capita GDP was 20% of the United States'; in 2000, the figure was 88%.[1] The performance of Hong Kong mirrors that of Singapore's, and the performance of China is no less astounding. Between 1978 and 2000, China's per capita GDP grew at an average of 8% per year. Resurgent Asia shows that backwardness does not have to be permanent, and that the thrust of divergence can be arrested and reversed—provided that the conditions are right.

What are those conditions, and how can a country get them right? There is a sizable body of literature on convergence and conditional convergence (Barro and Sala-i-Martin 1992, 1995; Baumol 1986; Baumol and Wolff 1988; DeLong 1988; Dowrick and Nguyen 1989; Mankiw et. al. 1992; Quah 1993, 1996; Pritchett 1995, 1997). I have no intention to enter into the debate on whether the convergence rate is greater or less than 2% per year. In Western Europe, North America, and Resurgent Asia, history has shown that innovation and diffusion of technology across countries hold the key to economic development.[2] The development of European countries—especially France, Germany and Italy—and of the United States owes much to the technological and commercial spillovers from Britain's great industrial power. The development of Japan and Resurgent Asia, in turn, owes much to the technology transfer from Europe and the United States. To be sure, the high savings rate of Singapore buttresses her growth, and the huge post-WWII capital inflow to Japan spurts Japanese development,

---

[1] The figures are extracted from the World Bank WDI *CD-ROM* (2002). The series used is GDP per capita at constant 1995 prices.

[2] According to Coe and Helpman (1995), industrialized nations benefit significantly from each other's R&D spillovers. For LDCs, international R&D spillovers must be even more important since LDCs spend little on R&Ds.

but it is inconceivable that either Singapore or Japan could have succeeded without the blessing of large scale technology transfer from abroad. The model in this paper focuses on the mechanism of technology transfer in order to understand income divergence among countries.

The picture of world growth and broad divergence almost conceals the picture of what Quah (1996) calls "twin peaks" group-convergence. As Resurgent Asia converges towards the affluent West, the African and other poor nations converge towards stagnation and starvation. In 1960 the Sub-Saharan group's per capita GDP was 3.6% of the United States' (US$473), and in 2000 the figure was 1.7% (US$564). For the poorest of the poor, 42 countries were designated by the World Bank as "Heavily Indebted Poor Countries – HIPC," whose per capita GDP in 2000 was only US$321. It is pointless to blame these poor countries for not saving and investing enough from their income of US$0.88 per day. To be sure, countries in this stagnant club have dictatorial governments, corrupt officials, and lethargic bureaucracies, but such factors are outside the realm of economic policies. What about technology? How does Resurgent Asia transfer technologies so successfully? Why has the transfer of ideas not been equally successful in poor countries?[3]

According to Romer (1993), countries are poor because they lack physical capital, human capital, and ideas. He stresses the distinction between human capital and ideas. Romer argues that to grow is to bridge the object (physical and human capital) gap as well as the idea gap. More specifically, he writes, "There are no easy solutions to an object gap. Objects have opportunity costs. . . . Idea gaps, in contrast, are relatively easy to solve. Ideas do not have opportunity costs" (1993; p.545-46). When a country is poor, it is hard to get them to invest in physical and human capital. However, it is easy, according to Romer, to at least tell them what advanced technologies are available so that they can use them accordingly (in his example, just to tell our friend that the file-recovery software exists).

The puzzle is that if ideas are so easily transferred, why have they not ameliorated the fate of the poor countries? Certainly the stock of knowledge and ideas has flourished and multiplied during the second half of the past century, but one hardly needs to be persuaded that the idea gap between the United States and poor Africa has grown dramatically. Romer's idea gap is strikingly similar to Gerschenkron's (1962) "relative backwardness," but Gerschenkron went even further than Romer did. According to Gerschenkron's well known "relative backwardness hypothesis," "... the opportunities inherent in industrialization may be said to vary directly with the backwardness of the country. Industrialization always seems the more promising the greater the backlog of technological innovations which the backward country could take over from the more advanced country" (1962; p.8). By this theory, catching-up should have become easier for Africa over the years since her idea gap has widened. The fact that Africa has fallen behind, and the fact that the world income distribution has diverged and become twin-peaked, suggests strongly that something is amiss in Gerschenkron's hypothesis.

[3] A group of theoretical models (Becker, Murphy and Tamura 1990; Azariadis and Drazen 1990; Galor and Zeira 1993; Galor 1996; Galor and Weil 1996) explains convergence clubs as multiple equilibria arising from internal income distribution, fertility, market imperfections, externalities, and other non-convexities. However, they share with the conditional convergence models in bypassing international technology interactions. An exception is Howitt and Mayer-Foulkes (2002), who study technology interactions in terms of innovation and diffusion that are, in turn, conditioned by institutions. In contrast to this paper, Howitt and Mayer-Foulkes do not focus on the learning mechanism.

I assert in this paper that the transfer of technology, of knowledge, and even of ideas is never automatic; it is never free. Three preliminary points immediately support this claim. First, it is no use giving a leading text in management to a less developed country (LDC) businessman and say, "just follow that." By the same token, it is no use telling a poor country to "follow the Korean model," or an impoverished athlete to "follow the Michael Jordon model" (see Lucas; 1993, p.252). Second, productive knowledge is embodied in business experiences of successful entrepreneurs, and in the workers they employ. To transfer these useful ideas overseas, businesses have to make foreign direct investment (FDI) and send their foremen and technicians abroad. Because ideas are embodied, ". . . for the poorest developing nations, letting multinational firms profit from the international transmission of ideas is the quickest and most reliable way to reduce the idea gaps that keep them poor" (Romer, 1993; p.548). What Romer does not explain, however, is why even multinational firms do not always transfer their technology. Third, because knowledge is embodied, technology transfer is a complex process of learning. To help the poorest developing nations, it is not good enough for multinational firms to operate and earn profits there. As Romer puts it, they have to profit from transmitting ideas, from transferring the technology, and from teaching the indigenous population how businesses are done. Even when the teachers are willing to teach and the students are eager to learn, transmitting ideas is far from easy or automatic.[4]

The thesis of this paper is as follows. Learning, the process by which knowledge is transferred from technology leader countries to follower countries, is a central feature of world development. Gerschenkron's "relative backwardness" thesis reveals half of the learning mechanism. Only by constructing the second half, will we be able to explain growth-divergence, twin peaks, and the determination of convergence-club memberships.

The plan of this paper is as follows. Section II reconstructs the "relative backwardness" learning mechanism, associates it with the analytical framework of Nelson and Phelps (1966), and upgrades it to a complete learning model. Growth-divergence and club-convergence are both explained in Section II. Section III extends the model to study the impacts of education on growth. Section IV searches for escape routes from the zero-growth club, and Section V summarizes and concludes the paper.

II. Two Models of Learning

Although never explicitly stated as such, Gerschenkron's "relative backwardness" thesis implies a linear model of learning. Sub-section II.A below substantiates both this assertion and the claim I made in Section I that Gerschenkron's model reveals only half of the learning mechanism. Sub-section II.B appends the other half and develops a complete learning apparatus, which is the basic model of this paper. The primary results of this paper will be developed in Subsection II.B.

A. "Relative Backwardness"

Emboldened by what he saw as the Soviet industrialization success, Gerschenkron (1962) conjectures that when a country is backward, it is blessed by the backlog of technologies that can be taken over from the more advanced countries. He writes, "Borrowed technology, so much and so rightly stressed by Veblen, was one of the primary factors assuring a high speed

[4] For the successes and failures of FDI to transfer technology see Görg and Strobl (2001) and the references therein.

of development in a backward country" (Gerschenkron; 1962, p.8). In another passage of his quoted earlier in Section I, he argues that the opportunities in industrialization vary directly with the backwardness of the country. Notice the implied monotonicity: the greater the backwardness, the more things there are to learn, and the faster one can master the backlogged technologies. This is contrary to the conventional, gradual approach of learning. All schooling systems, for instance, are gradated. Students move from primary schools, to secondary schools, and then to universities. Because knowledge is cumulative, each grade provides students with the capacity to benefit from the next grade. For example, if a seventh grader were put through a university course on, say, intertemporal optimization, he will no doubt find a huge amount of "backlog" material in front of him. He is unlikely, though, to retain much by the end of the course. Nevertheless, let us begin by formalizing this monotonic learning mechanism of "relative backwardness." Doing so helps expose its shortcomings, and helps unveil the upgrades needed to build an improved learning mechanism.

For the purpose of this paper, I keep production very much in the background. Let the aggregate production of a small backward country be $A(t)\cdot F\left[K(t),h(t),L(t)\right]$, where the variables from left to right are the Hick-neutral technology, physical capital, human capital, and labor at time t. The economy grows along the balanced path as described by Solow (1956), and growth rate is equal to the rate of technical change, $\dot{A}(t)/A(t)$. $A(t)$ is the country's achieved technology, which is some distance behind the international state-of-the-art technology, denoted $T(t)$. The degree of backwardness is the technology gap, which is a term I have used interchangeably with the term idea gap, $\left[T(t)-A(t)\right]$. Gerschenkron's "relative backwardness" posits that the speed of change of $A(t)$, $\dot{A}(t)$, rises monotonically with the idea gap $\left[T(t)-A(t)\right]$. Replacing monotonicity with proportionality helps fix ideas, and loses no generality. Thus we have $\dot{A}(t)=\alpha\left[T(t)-A(t)\right]$. Following Nelson and Phelps (1966; p.73), I assume the degree of proportionality α to be a function of the country's human capital stock $h(t)$. Thus the Gerschenkron-Nelson-Phelps (G-N-P) learning equation takes the form

$$\frac{\dot{A}(t)}{A(t)}=\Phi(h)\cdot\left[\frac{T(t)-A(t)}{A(t)}\right],\quad \Phi(0)=0,\quad \Phi'(h)>0 \quad . \tag{1}$$

Although Nelson and Phelps never associate their work with Gerschenkron's, the association is clearly seen in (1). The key point Nelson and Phelps make in their paper is that "... education has a positive payoff only if the technology is always improving" (1966; p.70). This is a penetrating insight that has made (1) one of the foremost models of learning. For education to have a payoff, and for the economy to grow, the technology gap must remain open. This open gap is ensured by the relentless technological progress at the frontier. Following Nelson and Phelps, let $\dot{T}(t)/T(t)=\lambda>0$ and $T(0)=T_0$. Using this to solve the first-order difference equation (1) yields a long-term steady state depicted by point X in Figure 1.[5] To the right of $\lambda/\Phi(h)$, the idea gap is large so that $\dot{A}(t)/A(t) > \lambda$, which implies catching up, and the gap falls towards $\lambda/\Phi(h)$. To the left of $\lambda/\Phi(h)$, the idea gap is small so that $\dot{A}(t)/A(t) < \lambda$, which implies falling behind, and the gap increases towards $\lambda/\Phi(h)$.

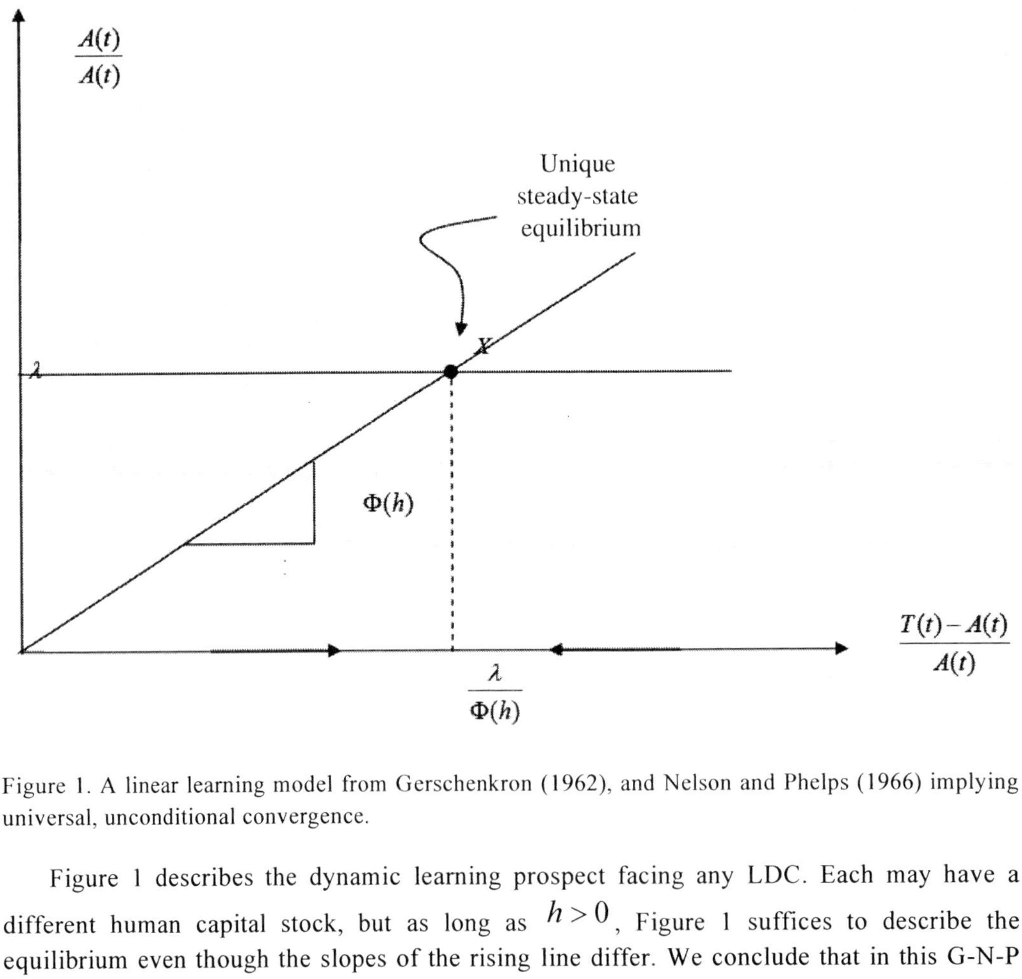

Figure 1. A linear learning model from Gerschenkron (1962), and Nelson and Phelps (1966) implying universal, unconditional convergence.

Figure 1 describes the dynamic learning prospect facing any LDC. Each may have a different human capital stock, but as long as $h > 0$, Figure 1 suffices to describe the equilibrium even though the slopes of the rising line differ. We conclude that in this G-N-P linear learning model, there should be growth convergence everywhere towards the rate λ, and that this must be true for any LDC with any human capital stock $h > 0$. A country with a

5 The explicit solution is $A^*(t) = \frac{\Phi(h)}{\Phi(h)+\lambda} \cdot T_0 e^{\lambda t}$, which can be rewritten in terms of an equilibrium technology gap $\frac{T(t) - A^*(t)}{A^*(t)} = \frac{\lambda}{\Phi(h)}$. For further details see Nelson and Phelps (1966).

high h will end up with a small, long-term technology gap, and a country with a low h will end up with a large, long-term technology gap. But for everyone, long-term growth rates will be λ. This is a strong result and though is harmonious with Gerschenkron's optimistic views about the Soviet economic history, it is unfortunately grossly at odds with the history of world development.[6] Growth-divergence is prevalent, yet "relative backwardness" precludes divergence. The G-N-P linear learning model cannot explain club-convergence, let alone the determination of club-memberships. Now the need for model upgrades is unmistakable.

B. Piece-wise Linear Learning

The most daunting difficulty G-N-P has with intuition is its monotonicity. Monotonicity assumes that the most backward countries have the best advantage for catching up. If this were true, Bangladesh and Chad should find it easier to learn advanced automotive technology than, say, Ireland or Singapore. This is counter intuitive. As we argued earlier, we expect learning to be gradual and cumulative. Ronald Findlay, in his study of FDI, finds it difficult to accept this idea of relative backwardness. He writes, "Of course, the disparity [*i.e.*, the technology gap] must not be too wide for the thesis to hold. Stone Age communities suddenly confronted with modern industrial civilization can only disintegrate or produce irrational cultural responses such as a 'cargo cult'" (1978; p.2).[7] Unfortunately, Findlay continues to follow the linear model despite the logical objection that he has just raised.

In order to build an improved learning model, there is no need to abandon relative backwardness and linearity altogether. Relative backwardness and the backlog of technology offers many advantages for the LDCs, but the technology gap "must not be too wide" for learning to take place. When this gap widens, absorption slackens, and though learning may continue, it is at a diminishing rate. If this gap continues to grow unhindered, learning will eventually come to a halt. The Stone Age community cannot comprehend Space Age technology.

---

6 For the historians, Gerschenkron's interpretation of the Soviet economic development has for some time been the subject of methodological attacks by scholars such as Barsby (1969), Sandberg (1982), and most fatally by Gregory (1984). Gregory argues that tsarist Russia was well on its way to economic development during 1885 to 1913. In addition, "Contrary to the official Soviet claims, real per capita income likely declined between 1913 and 1928" (Gregory; p.194).

7 The "cargo cult" is founded on a fallacious reasoning: *post hoc ergo propter hoc—after this therefore because of this*. The residents of Papua, Yaliwan, Vanuatu and other places noticed that when the colonial occupiers built wharves and airstrips, the wharves and airstrips were soon visited by ships and airplanes which delivered cargos of goods. They concluded that the ships and airplanes arrived as a consequence of the building of the wharves and airstrips, so they built their own wharves and airstrips in the expectation of receiving their own cargoes.

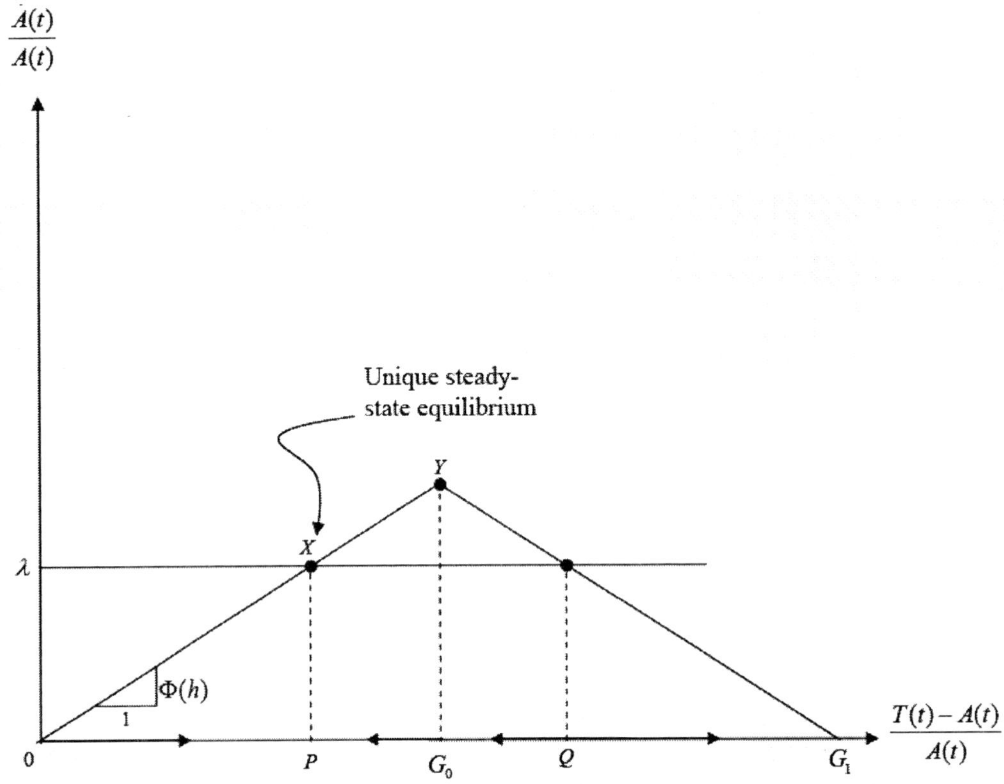

Figure 2. A preliminary piece-wise linear learning curve showing economic backwardness, equivalent to a widening technology gap, encourages learning, but only up to $0G_0$. Unlike G-N-P, this new model explains both growth-divergence and club-convergence.

As a preliminary exercise we retain the G-N-P linearity, up to a point *Y*, and append to it another linear, negatively-sloped segment YG_1 as shown in Figure 2. As long as the technology frontier progresses reasonably slowly, in particular $\lambda \leq YG_0$, point *X* remains the unique steady-state solution. The reason is as follows. First, countries with technology gap $\frac{[T(t)-A(t)]}{A(t)} \leq 0G_0$ will converge towards gap *P*. We saw this in subsection II.A. Second, countries with technology gap $0G_0 < \frac{[T(t)-A(t)]}{A(t)} \leq 0Q$ will also converge towards gap *P* because the principle of relative backwardness still applies. The backlog of technology is large but not too large, $\dot{A}(t)/A(t) > \lambda$, so catching-up continues and the gap shrinks. Third, countries with technology gap $0Q < \frac{[T(t)-A(t)]}{A(t)} \leq 0G_1$ will converge towards G_1 because the gap is too wide. As a result, there is a great deal to learn but little is retained, $\dot{A}(t)/A(t) < \lambda$, and learning diminishes until it collapses completely. Finally, countries with technology gap $\frac{[T(t)-A(t)]}{A(t)} > 0G_1$ will remain stagnant, and their distance from the frontier will continue to widen.

For a single LDC, given her capital stock h and exogenous λ, the future depends on her initial technology gap from the frontier. She will take the path to industrialization if this gap is sufficiently small, but she will take the path to stagnation if this gap is large. Now suppose Figure 2 depicts any LDC. Let them be identical in all respects except for their initial technology gaps from the frontier.[8] With identical learning curves, LDCs will only be distinguished by their positions along the horizontal axis of Figure 2.

Once we apply the reasoning presented earlier, the unique, long-term, steady-state solution is the point X given in Figure 2. Those countries located along the segment $0Q$ will receive membership invitations to join the λ-growth club, while those countries with an initial gap above $0Q$ will receive invitations to join the zero-growth club. This is our first model exhibiting both growth-divergence and club-convergence. In other theoretical models (see footnote 2 above), stagnation is the result of the LDCs' internal deficiencies – extreme income inequality, high fertility, and market failures. In our model, it could all be due to their initial idea gap being too wide; they could merely have started too late in history.

III. The Psychology of Learning

While the piece-wise linear learning curve is an improvement on the G-N-P model, the linearity is crude, and the abrupt reversal of slopes at the peak is arbitrary. More importantly, we need a thorough understanding on the learning mechanism in order to determine what parameters determine the shape and the position of the learning curve. Building a theoretical foundation for the learning curve is the motivation for this section.

Few scholars have spent more time studying the mechanics of learning than psychologists, and few learning theories are better known than the Yerkes-Dodson Law.[9] Yerkes and Dodson (1908) report that (a) mice learn tasks best at moderate levels of external stimulation, and (b) the more complex the task, the lower the level of optimal stimulation. Later researchers led by Hebb (1955) show that people perform best at a moderate level of arousal, with performance deteriorating under excessively high or low arousal levels. The Yerkes-Dodson Law is commonly depicted by an inverted U-shaped curve shown in Figure 3.

---

8 To avoid confusion, it is important to note that conceptually the position of one's learning curve as shown in Figure 2 depends also on the state-of-the-art technology. A more complex technology will be more difficult to learn and will imply a lower learning curve. We are, at this point, holding such complexity constant across LDCs, as well as holding λ constant. More on this in the next two sections.

9 I choose to focus on the Yerkes-Dodson, among other competing psychological theories of learning, because Yerkes-Dodson provides a convenient interface (owing to its inverted U-shape curve, see below) between G-N-P and the psychology literature. The psychology of learning is under-utilized as a building block in modeling economic growth. For the psychology literature see Brenner (1999, pp.15-43).

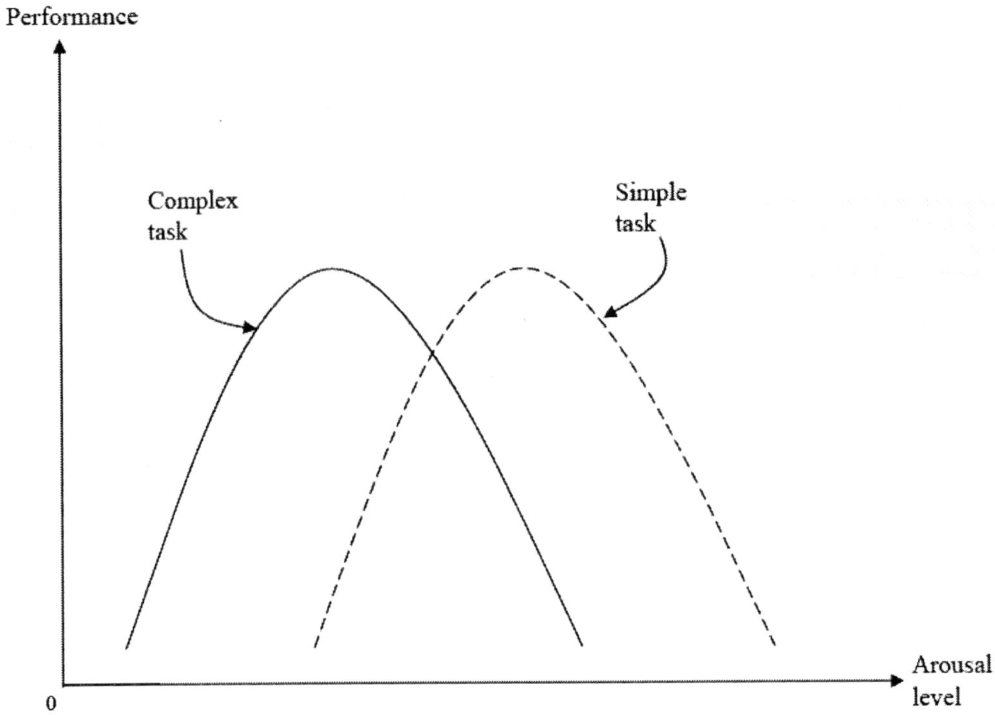

Figure 3. The Yerkes-Dodson Law states that the best performance occurs at a moderate level of arousal. Performance declines when arousal is below or above that level. The optimal level of arousal is lower for complex tasks (the continuous curve) than for simple tasks (the broken curve). This forms the basis for our new learning curve.

Arousal is commonly interpreted as stress, which Yerkes and Dodson deliver to their subjects via electric shocks of various intensities. In natural environments, students are stressed during examinations, and athletes are stressed during competitions. Kahneman (1973) argues that stress affects attention, and attention is crucial to learning. At low levels of stress we are bored and our minds stray. At high levels of stress we are "choked" and our minds vanquished. Neither of these situations is optimal for learning.[10] Now I want to postulate that when a LDC's technology gap increases, her citizens, grappling with increasingly novel environments, will likewise experience heightened degrees of stress. In other words I want to postulate an inverted U-shaped learning curve using the axes of Figure 4. Performance, taken as the speed of learning, is plotted on the vertical axis, and technology gap is plotted on the horizontal axis. When the gap is small, there is too little to learn and $\dot{A}/A$ is low; people are receiving insufficient stimuli to learn well. When the gap is large, there is too much to learn and again $\dot{A}/A$ is low; people are overloaded with information and they fail to pay attention to the relevant signals. When this happens, misguided associations

[10] Athletes who fail to perform up to a former standard are said to be "choked" from excessive anxiety. Students and learners suffer from information overload, and a massive gulf between their received knowledge and what they are put to learn could lead to "choking."

are made between the airstrips and the airplanes as in Findlay's example of "cargo cults." The learning curve in Figure 4 can be represented by

$$\frac{\dot{A}(t)}{A(t)} = \Phi(h) \cdot \Pi\left[\frac{T(t)-A(t)}{A(t)}\right]; \quad \Pi[\cdot] \geq 0, \Pi'[\cdot] \gtreqless 0 \text{ for } \left[\frac{T(t)-A(t)}{A(t)}\right] \lesseqgtr G_0, \text{ and } \Pi''[\cdot] < 0 \quad (2)$$

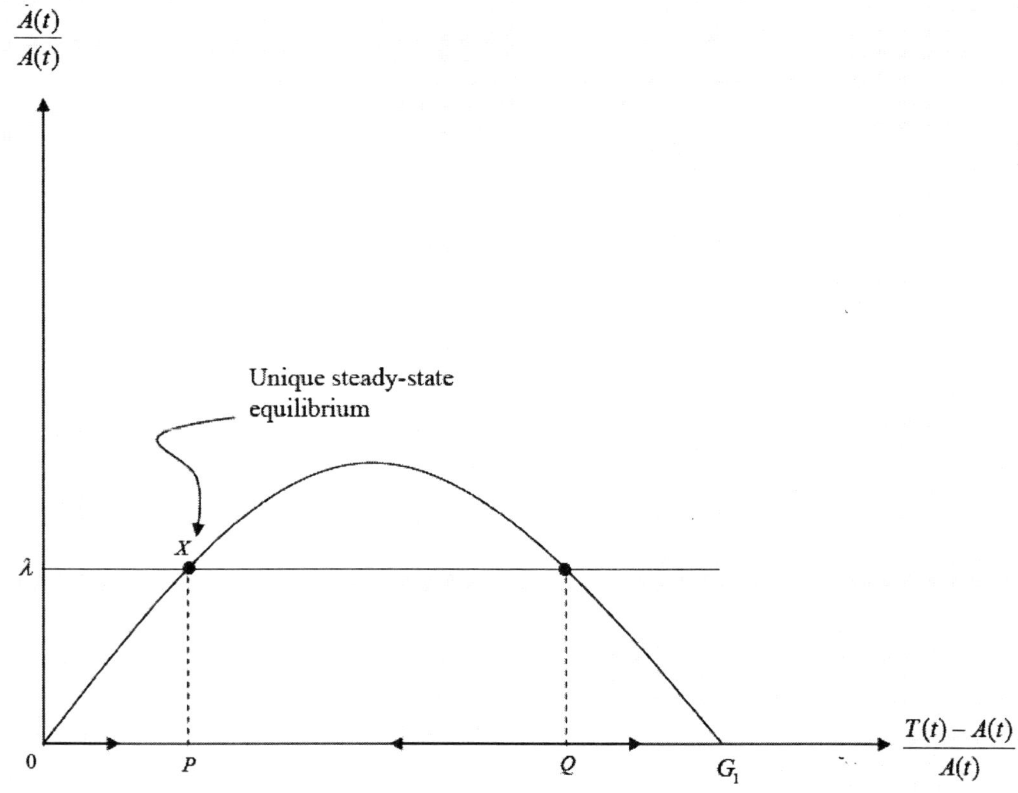

Figure 4. Given initial technology differences, the mechanics of learning explain the origins of growth-divergence and club-convergence.

Recall from Figure 3 that a learning curve represents a task of a particular degree of complexity. Suppose the task depicted in Figure 4 is the state-of-the-art (hence complex) technology which we referred to in Section II. Suppose also that this technology progresses at the speed λ. To establish that point X in Figure 4 is the unique, long-term equilibrium, all we have to do is repeat the reasoning used in Section II. Countries with an initial idea gap smaller than $0G_1$ will converge to growth rate λ, while countries with an initial idea gap larger than $0G_1$ will converge to growth rate zero. If all countries in the world have initial technology gaps smaller than $0G_1$, all will converge to the λ-growth club, and the zero-growth club will be empty. This is the policy-target that Section IV below seeks.

The literature on the psychology discussed in this section institutes a theoretical foundation of learning. We will now deploy this inverted U-shaped learning curve to continue with our investigations.

IV. The Origins of Growth-Divergence and Club-Divergence

Before tackling policy issues, important events in history need to be explained. The hallmark of the industrial revolution since 1820, as Maddison (2000) demonstrates, is to turn a "slow crawl" of the previous epoch into dynamic growth. The technology frontier, which Great Britain took over from the Dutch during the 19th century, and which the United States took over from the British during the 20th century, has been surging ahead at a much faster pace than ever before.[11] The fact that divergence did not occur during the 1000-1820 period of "slow crawl," and the fact that rapid industrialization, growth-divergence and club-convergence all took place during 1820-2000, are hardly a coincidence. The purpose of this section is to show that we can use our new learning curve to explain the origins of growth-divergence and club-convergence.

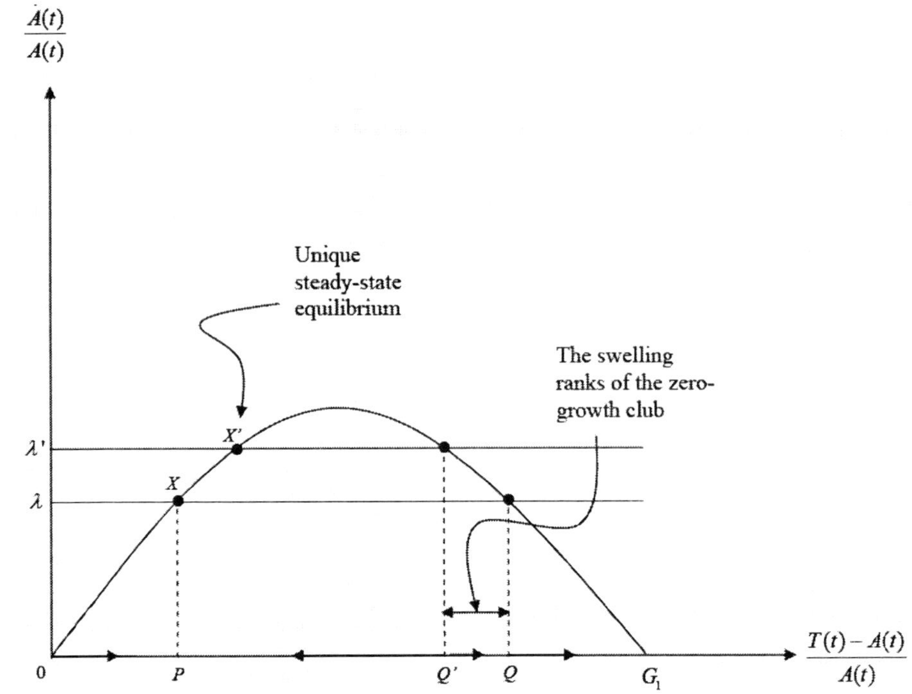

Figure 5. The origins of growth-divergence and club-convergence: industrialization since 1820 raises the exclusivity of the high-growth club and swells the ranks of the zero-growth club.

---

[11] Evidently, the past century's technological progress pales in comparison with that of our digital age. It is tempting to conjecture what this will mean for the future of growth-divergence and club-convergence. I will return to this conjectural exercise in the concluding Section VII.

In Figure 5, Western industrialization is heralded by an uplift of the horizontal line λ to λ'. A group of LDCs with initial technology gaps between Q' and Q in Figure 5, more precisely $0Q < \frac{[T(t) - A(t)]}{A(t)} < 0Q'$, will be relegated to the zero-growth club. Furthermore, the higher λ' is, the more exclusive memberships to the industrialized-club will be. It is as if advanced economic powers now immiserizes other nations not by colonialism or slavery, but by simply running fast, and inadvertently shaking off the followers at the end of the pack.

V. The Role of Education

So far, we have treated human capital h as exogenous and unchanging. h rises with education, and in this section we discuss education's impacts on our results. Our analysis will be conducted with the non-linear learning curve depicted in Figure 6.

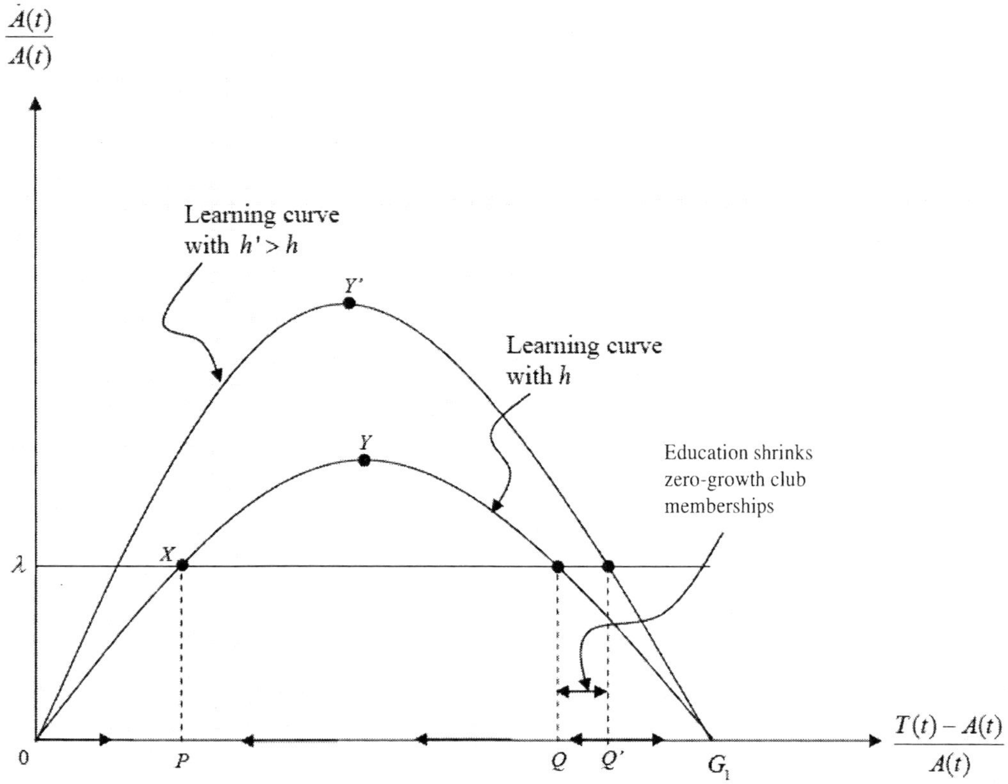

Figure 6. Educational effort and higher h shifts the learning curve upwards, but the two curves never cross. An LDC situated within QQ' switches to the λ-growth club. In general, the λ-growth club swells in size and the zero-growth club shrinks.

Suppose a single LDC raises her educational effort, which raises h. This shifts every point on the learning curve $0YG_1$ proportionately upwards to $0Y'G_1$ in Figure 6. Notice that the upper learning curve cannot cross the lower one otherwise a more educated workforce will learn less well, which is inconceivable if education is meaningful. This is confirmed by the fact that $\Phi(h)$ is a shift-parameter in the learning curve (2).

If her gap is between Q and Q', or $0Q < \frac{[T(t) - A(t)]}{A(t)} < 0Q'$, the LDC will leave the zero-growth club and join the λ-growth club. If her gap is outside this QQ' range, her educational efforts will not be reflected in growth rate change at all. This result helps us to understand Pritchett's (1996) finding on the inefficacy of education in LDC growth. Let me spell this out more precisely.

Suppose that all LDCs' learning curves are alike, and that this is true both before and after they raise h to h'. All LDCs within the QQ' range leave the zero-growth club to join the λ-growth club, and all other LDCs keep their old club-memberships unchanged. The proportion of countries that change membership, however, may be small. How small this proportion is depends on the extent of the learning curve shift, and on the existing distribution of LDCs along the horizontal axis of Figure 6. Suppose the shift in the learning curve is as shown, and the existing distribution is uniform along the domain $0G_1$. The proportion of LDCs that change memberships is equal to only $QQ'/0G_1$. The proportion of the poorest LDCs who reverse their fortunes, however, is QQ'/QG_1, and need not be small.

Pritchett (1996) finds that education has little effect on the growth rate of output per worker across a wide spectrum of countries. He offers three possible explanations: (1) schooling creates no human capital, (2) the marginal returns to education are falling rapidly, and (3) the institutional environments in many LDCs are perverse. All these may very well be true, but our result adds a forlorn synopsis: the poorest of the poor may have already moved down to the region between Q' and G_1. Help has arrived too late as they are already out of reach as far as additional education is concerned. Looking at divergence in this way adds a time dimension to remedial policies: education helps growth only if the timing is right. This conclusion implies, indirectly, that different stages of development require different educational institutions. While Korea and Singapore need to build world-class universities to compete in cutting-edge research, Chad and Ethiopia need basic literacy for its workforce. It seems improbable that all countries, irrespective of their stages of development, need the same kinds of schools. More work is needed on the relations between different type of schools and the different stages of economic development.[12]

---

12 The World Bank prefers to adopt a broader-stoke policy. Henry Rosovsky, Co-Chair of the Bank's 2000 Task Force on Higher Education and Society, says, "Higher education is the modern world's 'basic education,' but developing countries are falling further and further behind. It's time to drive home a new message: higher education is no longer a luxury, it is essential to survival." Rosovsky seems to be arguing that higher education is needed by all countries, to the same extent, and across the board. Our model implies, on the other hand, that backward countries need higher education to a lesser extent than the DCs. Rosovsky's quote is taken from the following website: (http://publications.worldbank.org/ecommerce/catalog/product?item_id=212746)

VI. Escapes from the Zero-Growth Club

As Maddison (2001) shows, the zero-growth club is not empty; it has been recruiting new members in the past century. It will be immensely satisfying if some escape routes are found, so that all members of the zero-growth club can be welcomed into the λ-growth community. If the predicaments of those in the zero-growth club are due, as the multiple-equilibria models suggest, to market imperfections, externalities and other non-convexities, then policies should be targeted at those non-convexities. But if they are due, as our model here suggests, to the learning dynamics, a completely different set of instruments are necessary. Since zero-growth memberships are issued to countries with learning difficulties, these slow learners should find a sufficiently simple technology from which they can comfortably learn. Within our framework of learning, there are two ways to characterize a technology—its complexity and its rate of progress (λ). Each characteristic gives rise to an escape route from the zero-growth club, and they are discussed in the two following subsections.

A. Choosing a Simple Task

Recall from the Yerkes-Dodson Law and from Figure 3 that the optimal level of arousal is lower for complex tasks than for simple tasks. Denote the complex task by T and the simple task by T'. Difficulties in learning from T reflect failures in disentangling the large amount of cues associated with T. By turning to T', learning improves because the amount of cues is lessened.

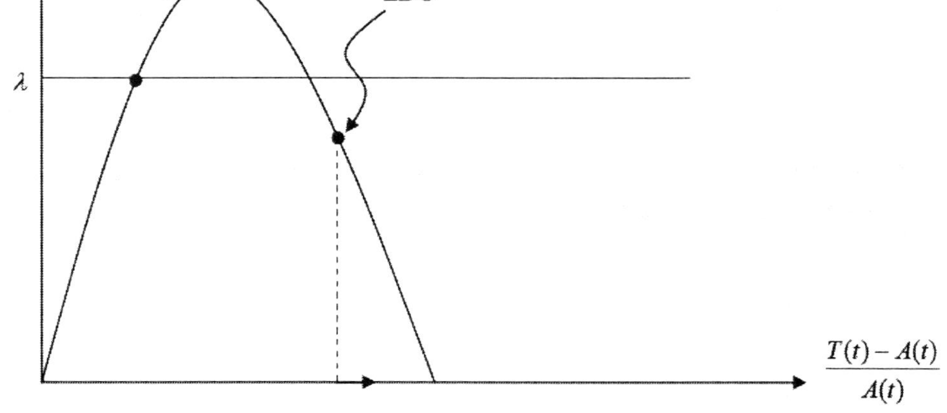

Figure 7.1. Learning from a complex task: state-of-the-art T leads to zero-growth membership.

Suppose the complex task, T, is the state-of-the-art technology. Suppose also that both T and T' progress at the same rate $\dot{T}/T = \dot{T}'/T' = \lambda$ (we relax this assumption in Subsection B below). Figure 7.1 shows the LDC belonging to the zero-growth club if confronted with the state-of-the-art T. However, by adopting T', as shown in Figure 7.2, she will instead join the λ-growth club. Technology T', if it exists, is her appropriate technology.

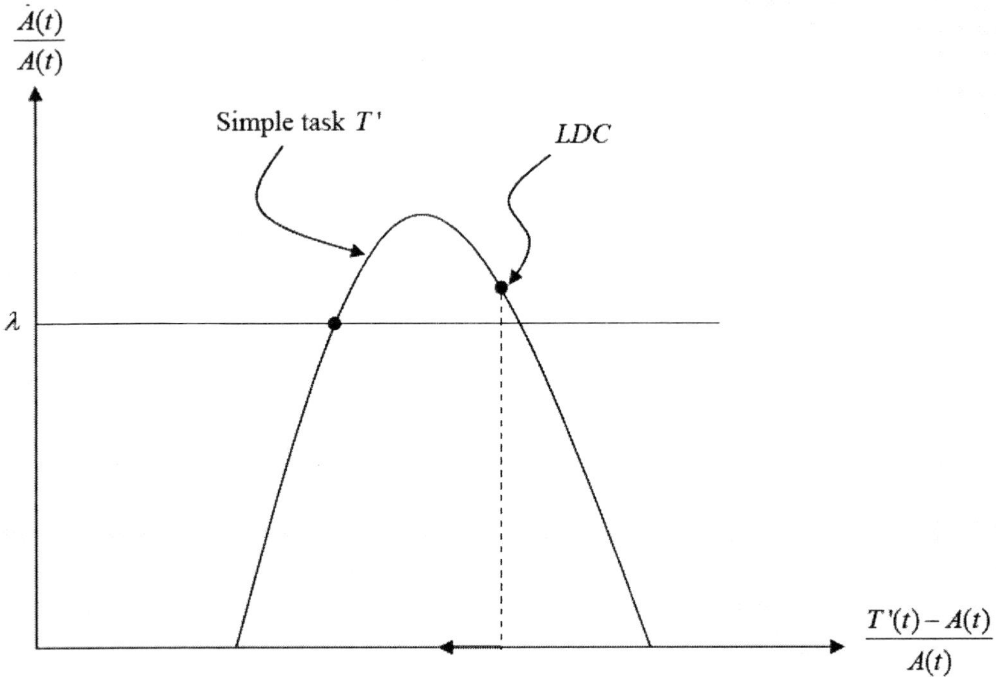

Figure 7.2. Learning from a simple task: T' leads to λ-growth membership.

While there usually is a unique, state-of-the-art technology to accomplish a task at any time t, there are usually "other ways to skin the cat." Suppose a sufficiently fine gradation of technologies exists between T and T' in terms of their degrees of complexity. Having ridden through the learning curve to the left in Figure 7.2, a country will have improved her technology A, and will be ready to tackle a more complex technology than T'.

As mentioned above, this learning process is comparable to schooling, where a student ascends from one grade to the next as he moves towards eventually graduation. For a LDC, there is a whole spectrum of technologies besides the state-of-the-art T that are practiced by a gradation of nations at various stages of development. For instance, when China began to open her doors in the early 1980s, the majority of foreign investors were Hong Kong businessmen who relocated their labor-intensive, simple processing and assembly operations across the border.[13] China did not get off the economic starting block by chasing the American and the Japanese state-of-the-art technology—though this was partly not by choice.

[13] See Thoburn et. al. (1990).

The high-tech U.S. and Japanese multinationals were not interested in China until quite recently.[14] In stark contrast to Gerschenkron's theory, our recommendation is this: The higher the degree of backwardness, the more elementary the class LDCs should enroll into.[15]

B. Choosing a Slower-moving Target

To develop our second and final escape route, suppose instead all tasks have the same complexity, but some progress with a more rapid pace than the rest. In other words, we assume a range of technologies $(T...T')$, differentiated not by their degree of complexity, but by their rates of progress $(\lambda < ... < \lambda')$. As Nelson and Phelps (1966) point out, learning ceases in the long run if technology is stagnant. Nelson and Phelps also imply, however, that learning is in essence shooting a target that moves. A toddler must first learn to walk before he can run; a bullock-cart community must first learn how to make bicycles before she sets her sights on, say, the maglev train. These points are made more concrete in Figures 8.1 and 8.2, which depict technologies T and T' respectively, which move at rates $\dot{T}(t)/T(t) = \lambda > \lambda' = \dot{T}'(t)/T'(t)$.

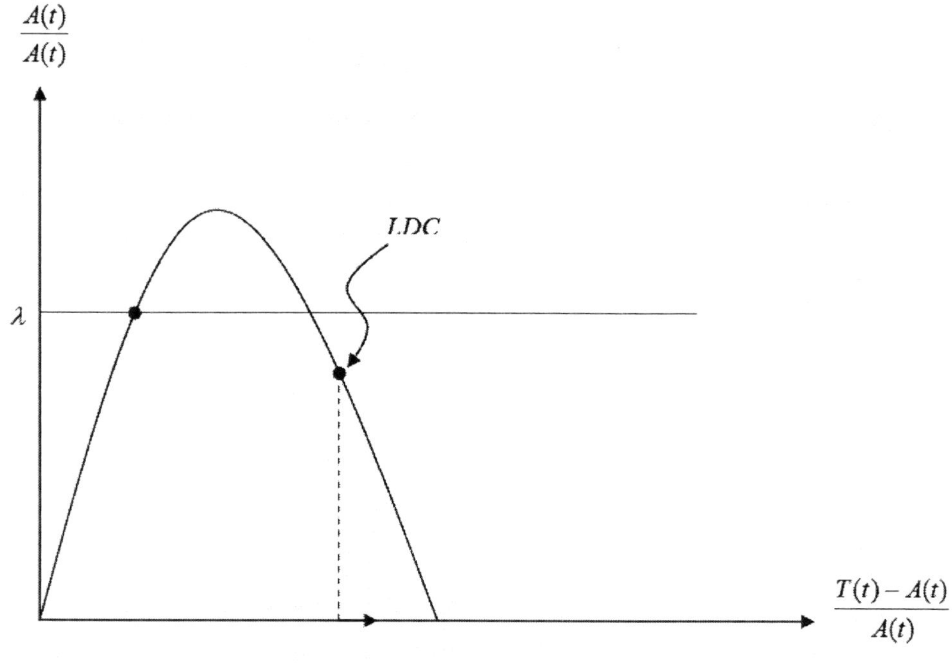

Figure 8.1. Learning from a fast-moving target: state-of-the-art T leads to the zero-growth club.

14 Lucas (1990) discusses "why doesn't capital flow from rich to poor countries."

15 Gerschenkron, on the other hand, commended the Soviet growth strategies when he said, ". . . the tendencies in backward countries to concentrate much of their efforts on introduction of the *most modern and expensive* technology, their stress on large-scale plant, and their interest in developing investment-goods industries need not necessarily be regarded as flowing mainly from a quest for prestige and from economic megalomania" (Gerschenkron, 1962; p.26, emphasis added).

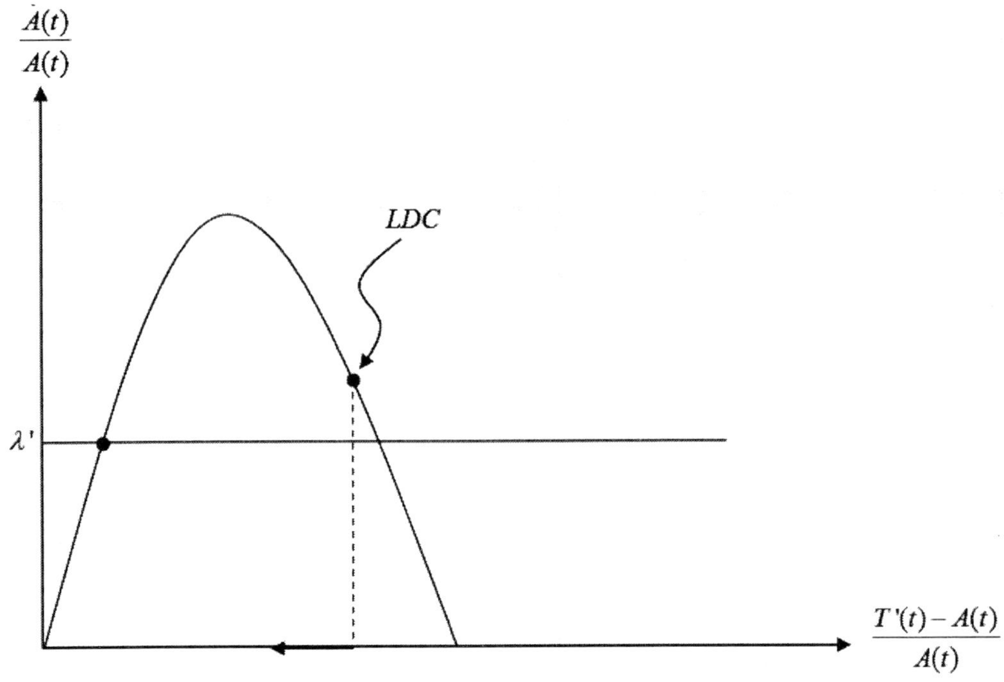

Figure 8.2. Learning from a slower-moving target: T' leads to the λ-growth club.

A LDC, with her backwardness, joins the zero-growth club if she adopts T (Figure 8.1), but joins the λ'-growth club if she adopts T' (Figure 8.2). The gradated nature of learning is vividly demonstrated once again. When a LDC has mastered the business of bicycle manufacturing, she can then enroll into the Suzuki motorcycles class, and then the Hyundai sedan class . . . ; she will finally graduate to compete with Toyota and General Motors on an equal footing. To return to the Chinese example again, we see that the labor-intensive, processing and assembly foreign direct investments in the 1980s have been gradually replaced by increasingly sophisticated technologies in the 1990s. The Chinese growth performance has become the envy of many LDCs.

VII. Conclusions

This paper has two themes. The first is that the history of world economic development evolves where economic followers learn from the leader. To understand how countries grow, we must study not only how they save, learn-by-doing and innovate, but we must also study how they learn from the industrial leaders of the world. Easterly and Levine (2001) suggest that the attentions of growth theorists should move away from factor accumulation. We argue in this paper that the mechanism of international learning is a logical place to recommence our quest for growth.

Our second theme is that learning is a gradual and cumulative undertaking. Gerschenkron's (1962) "relative backwardness" thesis, on the other hand, argues that backward countries such as the ex Soviet Union must buy the "most modern and expensive" technology and jump to the cutting-edge technological frontier. Even though I disagree with Gerschenkron, it is from his astute insight on technology "backlogs" and Nelson and Phelp's skillful manipulation of the linear learning dynamics that an improved and upgraded model of learning is built. We also make use of the psychologists' literature on learning. We replace the Gerschenkron-Nelson-Phelps linear learning with an inverted U-shaped curve, which is adapted from the famous Yerkes-Dodson Law (1908). This re-modeled learning mechanism allows us to see both clearer and farther than Gerschenkron, Nelson and Phelps could see previously. We have shown in Sections II and III that this new learning curve offers an explanation to both growth-divergence and club-convergence.

In Section IV we investigate the origins of growth-divergence and club-convergence. We find that the origin lies in a nation's initial technology gap. As long as technologies are not identical across countries, our theory predicts that international differences will be continually augmented, and that convergence-clubs will be formed. The origins of divergence, in other words, can be found in the mechanics of learning itself. Our theory accords well with economic history since 1820. Before 1820 little could be learned from outside, and countries did not differ much in terms of per capita incomes. After 1820 learning became critical, those who succeeded to learn surged ahead, while those who failed to learn lagged behind. Our theory complements the multiple-equilibria explanations of growth-divergence. The respective policy recommendations, however, differ sharply on many grounds. The multiple-equilibria story tells policy makers to focus on correcting market imperfections, on controlling birth rates, and on improving government efficiencies. All these are undeniably important issues for growth. We suggest, in addition, that LDCs should formulate their technology learning policies carefully—where to learn, and from which countries and which technology to learn. We warn, in particular, against the perils of the state-of-the-art technology. We recommend instead a step-wise, gradated approach to learn and to grow.

Our focus on learning leads us, in Section V, to examine education as a popular strategy for growth. Education enhances citizens' capacity to learn, but real learning occurs after graduation from schools—when citizens apply what they have learned to economic productive activities. Just as schools are gradated, to reflect on our theme once again, the technologies one chooses to assimilate and to adopt should be carefully gradated, too. Our formal model reveals that as far as long-term growth rates are concerned, education helps, if at all, only a small proportion of countries to switch from the zero-growth to the λ-growth club. We argue that this offers a rationale to Prichett's much debated missing link between education and growth. Moreover, LDC growth policies should be pursued with a poignant urgency. The poorest LDCs are quickly descending to the far right-hand side of their learning curves; they could become so rooted in the zero-growth club that they are beyond the help from education.

Despite the worsening odds that are stacked again us, Section VI submits two escape routes from the zero-growth club. Both routes arise from the theme—gradated and cumulative learning—that permeates our thesis. The first escape recommends the choice of a simple, appropriate technology—citizens in the Stone Age community need to graduate from the kindergarten, the primary school . . . before they enroll for university degrees.[16] The second

[16] I should hasten to add that I am in no way purporting that a backward LDC does not need any universities. She indeed has great demand for well-educated government bureaucrats, entrepreneurs, managers and technicians. However, it is unlikely that Chad, or even Malaysia, will need as many graduates per population as the DCs.

escape recommends the choice of a technology target that moves at a relatively slow pace—the LDC's own technology needs to be running at a fast enough speed before she can tussle with the brisk impulse of the Space Age.

We mention in Subsection III that modern technology, triggered by computers and information, has created a current that rushes forward at an unprecedented speed. It is important to reiterate the dynamic nature of learning as we end this paper. No matter where a LDC is economically, she must start learning now and without delay. As the philosopher Lao Tse says, "Learning is like rowing against a current—as soon as you stop, you drift backwards."

ACKNOWLEDGMENTS

I thank Paoli Chang, Malcolm Dowling, Hian Teck Hoon, Winston Koh, Denis Leung, Suzanne Leung, Bobby Mariano, Kirpal Singh, Swee Liang Tan and Yiu Kuen Tse for discussions and other help. I am particularly grateful to Dawn Dekle for guiding me to the psychology of learning. Any remaining errors are mine alone. Research support from the Wharton-SMU Research Centre is gratefully acknowledged.

REFERENCES

Azariadis, Costas and Allan Drazen, "Threshold Externalities in Economic Development," *Q.J.E.* **105** (May 1990): 501-526.

Barro, Robert J. and Xavier Sala-i-Martin, "Convergence," *J.P.E.* 100 (Apr. 1992): 223-251.

_____, *Economic Growth*, McGraw-Hill, Inc., 1995.

Barsby, Steven L., "Economic Backwardness and the Characteristics of Development," *J. Econ. History* **29** (Sep. 1969): 449-472.

Baumol, William J., "Productivity Growth, Convergence, and Welfare: What the Long-Run Data Show," *A.E.R.* **76** (Dec. 1986): 1072-1085.

Baumol, William J. and Edward N. Wolff, "Productivity Growth, Convergence, and Welfare: Reply," *A.E.R.* **78** (Dec. 1988): 1155-1159.

Becker, Gary S., Kevin M. Murphy and Robert Tamura, "Human Capital, Fertility, and Economic Growth," *J.P.E.* **98** (Oct.1990): S12-S37.

Brenner, Thomas, *Modelling Learning in Economics,* Edward Elgar, 1999.

Coe, David T. and Elhanan Helpman, "International R&D Spillovers," *European Econ. Rev.* **39** (May 1995): 859-887.

De Long, Bradford J., "Productivity Growth, Convergence, and Welfare: Comment," *A.E.R.* **78** (Dec. 1988): 1138-1154.

Dowrick, Steve and Norman Gemmell, "Industrialisation, Catching Up and Economic Growth: A Comparative Study across the World's Capitalist Economies," *Econ. J.* **101** (Mar.1991): 263-275.

Easterly, William and Ross Levine, "It's Not Factor Accumulation: Stylized Facts and Growth Models," *World Bank Econ. Rev.* **15** (Aug. 2001): 177-219.

Findlay, Ronald, "Relative Backwardness, Direct Foreign Investment, and the Transfer of Technology: A Simple Dynamic Model," *Q.J.E.* **92** (Feb. 1978): 1-16.

Galor, Oded, "Convergence? Inferences from Theoretical Models," *Econ. J.* **106** (Jul. 1996): 1056-1069.

Galor, Oded and David N. Weil, "The Gender Gap, Fertility, and Growth," *A.E.R.* **86** (Jun. 1996): 374-387.

Galor, Oded and Joseph Zeira, "Income Distribution and Macroeconomics," *Rev. Econ. Studies* **60** (Jan. 1993): 35-52.

Gerschenkron, Alexander, *Economic Backwardness in Historical Perspective*, Cambridge: Harvard University Press, 1962.

Görg, Holger and Strobl, Eric, "Multinational Companies and Productivity Spillovers: A Meta-Analysis," *Econ. J.* **111** (Nov. 2001): F723-39.

Gregory, Pual R., *Russian National Income, 1885-1913*, Cambridge University Press: New York, 1982.

Hebb, Donald D., "Drives and CNS," *Psychological Rev.* **62** (Oct. 1955): 243-54.

Howitt, Peter and David Mayer-Foulkes, "R&D, Implementation and Stagnation: A Schumpeterian Theory of Convergence Clubs," paper presented at the Pan American Health Organization of the World Health Organization, 3-4 October, 2002.

Kahneman, Daniel, *Attention and Effort,* Englewood Cliffs, NJ: Prentice-Hall, 1973.

Lucas, Robert E. Jr., "Why Doesn't Capital Flow from Rich to Poor Countries?" *A.E.R.* **80** (May 1990): 92-96.

_____, "Making a Miracle," *Econometrica* **61** (Mar. 1993): 251-272.

Maddison, Angus, *The World Economy: A Millennial Perspective*, Development Centre of The Organization for Economic Co-operation and Development, 2002.

Mankiw, N. Gregory, David Romer, and David N. Weil, "A Contribution to the Empirics of Economic Growth," *Q.J.E.* **107** (May 1992): 407-437.

Nelson, Richard R. and Edmund S. Phelps, "Investment in Humans, Technological Diffusion, and Economic Growth," *A.E.R.* **56** (Mar. 1966): 69-75.

Pritchett, Lant, "Forget Convergence: Divergence Past, Present, and Future," *Finance and Development* (June 1996): 40-43.

_____, "Where Has All the Education Gone?" *World Bank Econ. Rev.* **15** (Oct. 2001): 367-391.

_____, "Divergence, Big Time," *J. Econ. Perspectives* **11** (Summer 1997): 3-17.

Quah, Danny T., "Empirical Cross-Section Dynamics in Economic Growth," *European Econ. Rev.* **37** (Apr. 1993): 426-434.

_____, "Twin Peaks: Growth and Convergence in Models of Distribution Dynamics," *Econ. J.* **106** (Jul. 1996): 1045-1055.

Romer, Paul, "Idea Gaps and Object Gaps in Economic Development," *J. Monetary Econ.* **32** (Dec. 1993): 543-573.

Sandberg, L.G., "Ignorance, Poverty and Economic Backwardness in the Early Stages of European Industrialization: Variations on Alexander Gerschenkron's Grand Theme," *J. European Econ. History* **11** (1982): 675-698.

Solow, Robert M., "A Contribution to the Theory of Economic Growth," *Q.J.E.* **70** (Feb. 1956): 65-94.

Thoburn, John, H.M. Leung, Esther Chau, and S.H.Tang, *Foreign Investment in China Under the Open Policy: The Experience of Hong Kong Companies,* Avebury: Gower Publishing Company, London, 1990.

World Bank Task Force on Higher Education and Society, "Higher Education in Developing Countries: Peril and Promise," 2000.

Yerkes, Robert M. and John D. Dodson, "The Relation of Strength of Stimulus to Rapidity of Habit-Formation," *J. of Comparative Neurology and Psychology* **18** (1908): 459-482.

In: Economics of Technological Change
Editor: Thomas L. Beane, pp. 23-38
ISBN: 978-1-60021-237-6
© 2007 Nova Science Publishers, Inc.

Chapter 2

INNOVATION, PATENT LIFETIME AND ECONOMIC GROWTH*

Alberto Bucci
University of Milan (Department of Economics), Milan, Italy
H. Cagri Saglam
Catholic University of Louvain (Department of Economics), Louvain-la-Neuve, Belgium

ABSTRACT

Since Schumpeter (1942), the need to provide the successful innovator with some form of patent protection in order to stimulate investment in R&D has been widely recognized. As patent protection helps innovators to benefit from their research efforts, it creates or increases the incentives to perform research activity. It also promotes technological progress and, thus, economic growth.

The analysis of optimal patent lifetime is founded on two branches of economic theory that in recent years have proceeded, for the most part, separately.

In the IO literature, the analysis is built on the hypothesis that a society must balance the gains accruing from rapid technological progress against the welfare losses associated with the presence of temporary monopolies in the use of a new technology (Nordhaus, 1969; Loury, 1979). However, this literature is mainly confined to study the incentives and the distortions induced by a patent system within a static, partial equilibrium framework. Judd (1985) represents the first attempt at studying the optimal patent lifetime through the maximization of a social welfare function in a dynamic, general equilibrium set-up. He builds an exogenous growth model where innovation, though endogenous, is not sustainable when there is no exogenous increase in the labor force. Still, his analysis remains important because it shows that patent lifetime can have important effects on the equilibrium innovation and growth rates.

Within the endogenous growth literature, innovation-based economic growth has become an important field of research due to the pioneeristic works of Romer (1990), Grossman and

---

* We thank B. Amable, R. Boucekkine, G. Cozzi, D. de la Croix and J. Ruiz for comments on a first draft of this chapter. We also wish to thank the participants at the Catholic University of Louvain Economic Theory Seminar, the 4th International Conference in Economics (Ankara, Turkey), the 29th EARIE Conference (Trinity College, Dublin, Ireland) for useful discussions. We are particularly indebted to Morton Kamien for his valuable insights on a previous version of this work. Usual disclaimers apply. Alberto Bucci acknowledges financial support from MIUR (*Italian Ministry for University and Scientific Research, grant number 2005131555_003*).

Helpman (1991) and Aghion and Howitt (1992). In these models, though, no attention is paid to the patent system as a *government policy tool* (it is assumed that a successful innovator always accrues an infinite patent protection against future innovators). More recently, Michel and Nyssen (1998) have analyzed the role of patent lifetime on growth in a similar framework. Nonetheless, their analysis remains limited as they consider a deterministic innovative activity. Loury (1979) was among the first to highlight the limits of modeling innovative activity as a deterministic process, instead of a stochastic one. On the other side, Aghion and Howitt (1992) and Grossman and Helpman (1991, Chapter 4) assume that innovative activity is stochastic but continue to maintain the assumption that patent life is infinite.

Our motivation in this chapter is to fill the literature gaps mentioned above. Using an endogenous technological change model with vertical differentiation and stochastic R&D activity, we address the following question: what is the patent lifetime a government would set in order to maximize aggregate economic growth? Within this framework we also study how this patent length would change in response to changes in the degree of competition in R&D, the productivity parameter of the innovation *production function* and the level of innovator's monopoly profits.

Keywords: Endogenous Growth; Product Differentiation; Innovation; Patent Lifetime

JEL Classification: O31, O38, O41

1. INTRODUCTION

A patent is an exclusive right to exploit (use, sell, or import) an invention over a limited period of time within the country where the application is made. The three most important characteristics that a new idea must have in order to be patented are *novelty* (it has to be original), *non-obviousness* (it has to be innovative) and *usefulness* (it must have an industrial application). Since Schumpeter (1942), the need to provide the successful innovator with some form of patent protection in order to stimulate investment in research and development activity (R&D, henceforth) has been widely recognized. As patent protection helps innovators to benefit from their research efforts, it creates or increases private incentives to innovate. It also seeks to avoid duplication of research efforts and to promote technological progress. On the other hand, however, patents have an important drawback: they reduce for a given period of time product market competition and the speed of technology diffusion. As a consequence, there is a need to ensure that the patent system strikes the right balance between the appropriation of the fruits of innovation by patent holders and the diffusion of technology for the society as a whole. Notwithstanding the recent trends of patenting activity across sectors and countries in the world and the crucial relationships that can take place among patents, social welfare, technological progress and economic growth, it is surprising how policy oriented research on such links has been until now relatively scarce (OECD, 2004a, p.11).

As an example, it is estimated[1] that between 1992 and 2002 there was a doubling of the number of patent applications at EPO (the European Patent Office). Among the EU countries, the leading ones in patenting activity over the second half of 90's have been Germany, Finland and Sweden. Across sectors, the most dynamic in patenting were biotechnology and ICT. The share of biotechnology in EPO filings climbed from 4.3% in 1994 to 5.5% in 2001. During the same period the share of ICT climbed from 28% to 35%. The increase in the EPO

[1] See OECD (2004b) for major details.

share of the countries mentioned above (Germany, Finland and Sweden) can be traced essentially to ICT patents. The ratio of patents to business-funded R&D (patents per dollar of R&D) has increased by 50% for European patentees at the EPO between 1994 and 2000 (such an increase was primarily due to Germany as a country and to ICT as an industry).[2] Behind this upsurge of patenting activity there can be different reasons. According to Eaton *et al.* (2004), about 60% of the nearly 6% annual increase in European patents between 1991 and 2000 can be attributed to the decline in the EPO application fees, whereas the remaining growth may be due to an increase in innovative output of around 2.3%. Second, patents are often employed as a strategic means of competition by firms. Indeed, today both large and small companies use patents not only for securing the technological space to develop new products, but also for blocking competitors in their development activities. Another, but not less important, reason for the surge of patenting activity is that patents are recognized as part of companies' intangible assets in the trading of knowledge and, as such, contribute importantly to foster firms' reputation. Finally, changes in patent regimes that have expanded the realm of patent protection and strengthened rights of patent holders have also played an important role in affecting the pace of patenting activity by firms around the world.

There are at least four dimensions of a patent system that policy makers could potentially use in order to enhance innovation and economic growth:

- *Patent Subject Matter*: is the domain of knowledge that can be patented, once a new idea has met the criteria of novelty, non-obviousness and usefulness. Scientific discoveries and/or too abstract ideas are, for example, generally not patentable. A precise definition of the patent subject matter may help to understand which discoveries really improve a society's welfare and, as such, merit (public) attention;

- *Patentability Requirement*: is a minimum innovation size required to receive a patent. A too low patentability requirement increases the risk of having many patented inventions with a very little social value. On the other hand, a too high patentability requirement increases the risk of discouraging those innovations that, although not radical, may importantly contribute to advance technological progress;

- *Lagging/Leading Patent Breadth*: Lagging breadth limits imitation by specifying inferior products that other firms cannot produce. Leading breadth limits future innovators by specifying superior products that other firms cannot produce. Both too broad and too narrow patents may discourage innovation;

- *Patent Life*: is the length of time for which a patent is valid. On the one hand, patent lifetime should be long enough to provide sufficient incentives to innovation; on the other hand, it should be short enough to allow for a proper diffusion of fundamental knowledge. There is no simple solution to this delicate balance.

---

[2] A similar pattern can also be documented for the United States, where the number of software-related patents has grown much faster than total patents granted in the near past, and now accounts for between 4% and 10% of all patents (OECD, 2002, p. 6).

Other two legal aspects of a patent that have a potential bearing on private incentives to innovate concern respectively the enforcement of the exclusive rights conferred to patent holders and the amount of damages attributed by courts in case of infringement. All together, these characteristics of a patent define the *strength of a patent system*.

Despite the efforts towards harmonization that have been made in recent years across countries (longer patent lifetime and expansion of the subject matter of patents, for example), the US and the European patent regimes continue to display important differences in two respects.[3] First of all, since the creation of a centralized court system in 1982 (the *Court of Appeal of the Federal Circuit*, CAFC), rights of patent holders are more frequently and strongly enforced in court in the United States, whereas the implementation of a centralized patent litigation system is still under discussion in Europe. At the same time, damage awards in patent litigation trials have substantially increased in recent years in the US. Secondly, the US patent system still appears as *more flexible* than the European one, as it allows *"...the final grant to be different (usually narrower) than from the initial application"* (OECD, 2004b, p.18). All this points to the need of increasing the harmonization efforts between these two international patent systems. It is widely believed that harmonization has to go especially in the direction of improving consistency of patent regimes and practices (evaluation procedures, standards and enforcement). Moreover, it is held that serious administrative reforms in Europe are needed in order to speed up patent awards and reduce patenting costs for small and medium-sized enterprises.

Until now economic theory did not pay the due level of attention to the analysis of the role patents may have in growth (Keely, 2001). With the arrival of new sectors (*e.g.* biotechnology and ICT) on the technological scene, economists and policy makers need to stimulate a deeper discussion on this important research topic.

2. Related Literature and Motivation

The analysis of optimal patent lifetime is founded on two branches of economic theory that, for the most part, have proceeded separately in recent years: the industrial organization and the endogenous growth literature.

In the *industrial organization theory*, the analysis is built on the hypothesis that a society must balance the gains accruing from rapid technological progress against the welfare losses associated with the presence of temporary monopolies in the use of a new technology. Following Nordhaus (1969) and Loury (1979), however, this literature is mainly confined to study the incentives and the distortions induced by a patent system within a static, partial equilibrium framework. Using a quality-ladder model, Horowitz and Lai (1996) assess the effects of patent length on the rate of innovation and consumer welfare. They find that the rate-of-innovation-maximizing patent length always exceeds the consumer-welfare-maximizing patent length. In their model, however, no explicit attention is paid to the growth-maximizing patent lifetime. Judd (1985) represents the first attempt at studying the optimal patent lifetime through the maximization of a social welfare function in a dynamic, general equilibrium set-up. He builds an exogenous growth model where innovation, though

[3] For a comparison between these two patent regimes, see, among others, the recent study by Martínez and Guellec (2004).

endogenous, is not sustainable when there is no exogenous increase in the labor force. Still, his analysis remains important because it clearly shows that patent lifetime can potentially have important effects on the equilibrium innovation and growth rates.[4]

Within the *endogenous growth literature*, innovation-based economic growth has become an important field of research due to the pioneeristic works of Romer (1990), Grossman and Helpman (1991) and Aghion and Howitt (1992). Considering innovations as a means able to create a market for differentiated products as a result of the increase in the degree of product variety or quality, in these models, however, no attention is paid to the patent system as a *government policy tool.* Indeed, within a model of horizontal product differentiation and growth, Romer (1990) and Grossman and Helpman (1991, Chapter 3) assume that a successful innovator accrues an infinite patent protection against future innovators. In a more recent work, Michel and Nyssen (1998) do analyze patent lifetime in a similar framework (with economic growth being induced by purposive R&D investment which aims at expanding the set of currently available horizontally differentiated goods). Though, their analysis is limited in the sense that they consider a deterministic innovative activity. Since Loury (1979), the limits of modeling innovative activity as a deterministic process, instead of a stochastic one, have been widely recognized. On the other side, Aghion and Howitt (1992) and Grossman and Helpman (1991, Chapter 4) assume that innovative activity is stochastic but maintain the assumption that patent life is infinite, in the sense that it is terminated only by future innovations. Following this last branch of the literature, O'Donoghue and Zweimüller (2004) examine in detail the role patent protection may play within an endogenous growth model with vertical differentiation. The main goal of their paper is indeed to illustrate how protection against future innovators can stimulate R&D relative to situations without such protection. At this aim, they formally introduce four tools of patent policy (*patent lifetime*, *a patentability requirement*, *lagging breadth* and *leading breadth*)[5] and focus on the simplest case where patent life is infinite (in the sense described above) and there is complete lagging breadth. Even though they do not address the question of the relationship between patent length and long-run economic growth their analysis suggests that general equilibrium models incorporating patents may produce results (regarding optimal patent policy) that are completely different from the ones obtained within a static, partial equilibrium context.[6]

Segerstrom *et al.* (1990) also examine a dynamic general equilibrium model with trade in which R&D activity and, hence, technological change are influenced by the length of patent protection and the height of tariffs. They find that increasing the length of patents in the North (the innovating region) can either increase or decrease R&D activity in that region. Although longer patents increase the returns to R&D, they may also imply that more fixed resources must be devoted to producing existing products, thus reducing growth.

---

[4] According to Judd (1985), if patent lifetime is infinite, then there will be either too much or too little innovation in equilibrium. On the other hand, if patent lifetime is finite, the equilibrium is not stable and cycles of innovation are generated.

[5] According to their definition, "...*Patent life is the length of time for which a patent is valid. A patentability requirement specifies a minimum innovation size required to patent a new product. A patent's breadth specifies products that would infringe upon the patent, which means that these products cannot be produced without the patent holder's permission (in the form of a license). Lagging breadth specifies inferior products that infringe, and leading breadth specifies superior products that infringe*" (O'Donoghue and Zweimüller, 2004, p.87).

Using a similar North-South model of trade, innovation, imitation and economic growth, Helpman (1993) concludes that the stronger the level of intellectual property rights, the less imitation there is in the South (the non-innovating region). He also finds that strong intellectual property rights will increase innovation in the short run as the profitability of innovation in the North increases. In the longer run, however, and similarly to what happens in Segerstrom *et al.* (1990), the rate of innovation actually falls because the North produces more technologically-old goods taking resources away from innovation. Again, both these two last papers are completely silent on the length of patents that guarantees the maximum level of economic growth.

Given all this, it is remarkable that the impact of patent lifetime as a government policy tool on economic growth is almost entirely absent from theoretical work. Our motivation in the present chapter is to fill this important and wide gap in the literature. At this aim, we use an endogenous technological change model with vertical differentiation and stochastic R&D activity in order to address the following question: What is the patent lifetime a government would set in order to maximize aggregate economic growth? Within this framework we also study how this patent length would change in response to changes in the level of competition in R&D (as measured by the number of firms performing research), the real interest rate, the productivity parameter of research technology and the monopoly profit accruing to the successful innovator.

We build upon the path-breaking Aghion and Howitt (1992) model of innovation and growth, with a major difference. Unlike their work, where, as already mentioned, a patent lifetime is potentially infinite (since it is cut only by a new quality improvement), we assume that it is government to determine the *legal* patent lifetime (T^*) in such a way that the amount of resources devoted to R&D by profit-seeking agents and, hence, aggregate economic growth are maximized. Through numerical simulations we show that T^* (the growth-maximizing patent lifetime set by government) increases with the level of competition in the R&D sector and the real interest rate, whereas it decreases with the monopoly profit of the successful innovator and the productivity parameter of research technology.

The chapter is organized as follows. Section 3 presents the model economy. Section 4 introduces economic growth. Section 5 characterizes, through numerical analysis, the growth maximizing patent lifetime and performs comparative statics exercises. Finally, section 6 concludes.

3. The Model

We consider an economy composed of three vertically integrated sectors. There is a final output sector that produces a homogenous consumption good, an intermediate sector that produces vertically differentiated capital inputs and finally a R&D sector producing innovations. Final output is obtained by combining labor and a capital good (available in different qualities) through a Cobb-Douglas technology and is used as an input in the intermediate and R&D sectors. An important aspect of the Schumpeterian growth model is that when an existing product is improved, the new good displaces the older one(s). In order

[6] Keely (2001) presents an extensive survey on the most recent attempts at introducing patents in dynamic general equilibrium models.

words, the different quality grades for a given good are close substitutes. We make the extreme hypothesis that the different qualities of the existing type of intermediate input are perfect substitutes (the discovery of a higher grade implies that the lower grades are driven completely out of the intermediate sector). This is the traditional *creative destruction effect* advocated by Schumpeter (1942). As a consequence of our assumption, in what follows we can focus on an equilibrium where only the leading-edge quality of the existing capital good is actually produced and, thus, employed by final output producers to generate the consumption good. Since the intermediate market structure is always monopolistic, the capital-goods producer sells his/her own product to final-goods firms at the monopoly price. Firms that operate in the R&D sector race to produce innovations in terms of quality improvements in the existing capital good product line. The successful innovator wins a patent that is licensed to intermediate producers. In order to engage in innovative activity R&D firms must bear a fixed cost (in terms of final output). The presence of such a fixed cost assures that at each time the number of R&D firms is bounded. In deciding how many resources to devote to research, R&D firms consider two variables: the size of the profit flow coming from this activity and its likely duration. This duration is clearly random, as it depends on the uncertain outcomes from the research efforts of competitors. A policy maker knows how many R&D firms are racing at each date for a new innovation and in a symmetric equilibrium (where R&D firms are all identical to each other) sets the patent lifetime in order to maximize the amount of resources devoted to innovation by the representative R&D firm and, then, the aggregate economic growth rate.

3.1. The Final Output Industry

Final output (the numeraire good in the model) is produced within a competitive industry and according to the following constant returns to scale production function:

$$Y_t = L^{1-\alpha}\left(\widetilde{X}_t\right)^{\alpha}, \qquad \widetilde{X}_t \equiv \sum_{m=0}^{M}\left(q^m X_{mt}\right), \qquad 0<\alpha<1, \qquad q>1, \tag{1}$$

where Y denotes total output of the consumption good and L is the fixed aggregate flow of labor supply. According to equation (1), output also depends on the input of an intermediate good, X, available in different quality levels. The potential grades of the intermediate good used into production of the homogeneous final output are arrayed along a quality ladder with rungs spaced proportionately at interval q>1.[7] X_{mt} is the quantity used (at time t) of the intermediate good with quality level m. With M being the highest quality level available for the intermediate input, $\widetilde{X}_t$ is its quality-adjusted amount employed at time t. α is a technological parameter strictly between 0 and 1.

Since we assume that only the top-quality generation (M) of the existing capital good product line is produced and used in equilibrium, the final output production function can be recast as follows:

$$Y_t = L^{1-\alpha}\left(q^M X_{M_t}\right)^{\alpha}. \tag{2}$$

[7] This set-up follows Aghion and Howitt (1992) and Grossman and Helpman (1991, Chap.4).

The representative firm producing the homogeneous final output maximizes its instantaneous profit with respect to X_M (from now on, we drop for ease of notation the subscript t on the variables depending on time):

$$\underset{X_M}{Max}\left[Y - wL - \left(P_M X_M\right)\right].$$

In equation (3), the price of the final good (Y) has been normalized to unity. P_M is the price of the highest grade of the intermediate good and w is the wage rate accruing to one unit of labor. The solution to this problem yields the demand schedule of the top-quality intermediate input (X_M):

$$X_M = \left(\frac{\alpha q^{\alpha M}}{P_M}\right)^{1/(1-\alpha)} L .$$

The implied price-demand elasticity (e) faced by the intermediate monopolist is given by the constant:

$$e = \frac{1}{1-\alpha}.$$

With $0 < \alpha < 1$, this elasticity is clearly greater than one.

3.2. The Intermediate Goods Sector

At any point in time, there is only one firm operating in this sector. This firm produces the highest quality intermediate good that is used in the production of final output as an input. One unit of foregone consumption allows the intermediate firm to produce one unit of intermediate good. Accordingly, the marginal cost (MC) beard by this firm is equal to one. On the other hand, its marginal revenue (MR) is given by:

$$MR = P_M\left(1 - \frac{1}{e}\right) = \alpha P_M .$$

In order to obtain the optimal price of the highest available quality of intermediate good the firm equates MC to MR. This leads to a price (P_M) equal to: $P_M = \frac{1}{\alpha}$. Thus, price is a fixed mark-up ($1/\alpha > 1$) on the marginal cost of production and is independent of the quality level currently produced (M). The intermediate firm, after having incurred the fixed-cost investment in a patent, will obtain the following instantaneous flow of profit:[8]

[8] At any point in time the only sunk cost faced by the firm operating in the intermediate sector is the initial expenditure on the patent.

$$\Pi_M = (P_M - 1)X_M = (1-\alpha)\alpha^{\frac{1+\alpha}{1-\alpha}} q^{\frac{\alpha}{1-\alpha}M} L \quad (5)$$

Note that, as long as L is constant, Π_M is constant, too. Moreover the profit turns out to be an increasing function of the quality rung, m. Thus, the profit received by the monopolist producing the higher quality intermediate good will be larger. Since in equilibrium q^m will increase over time, profits will also increase over time. A key difference between this class of models (*quality ladder models* of endogenous growth) and *expanding varieties* (horizontal differentiation) growth models is that in the present case the monopoly right to produce the top-quality intermediate input does not last forever, being cut by a new quality improvement made by a competitor. This is what we analyze more deeply in the next paragraph.

3.3. R&D Activity

In this section we postulate an R&D technology where the date of innovation is uncertain. Following Loury (1979) and Dasgupta and Stiglitz (1980), we assume a Poisson process for innovations that will bring discrete quality improvements to the intermediate good. The timing of innovations stochastically depends on the R&D effort chosen by each competitor, namely the amount of final output devoted to innovative activity, and is, therefore, endogenous. A firm introducing an innovation will maintain a legal patent which is licensed to an intermediate firm and whose lifetime is set optimally by government.

Concerning the structure of the sector, at time t each firm i determines the amount of resources devoted to R&D (N_i) through maximizing its own expected flow of profits (π_i) with respect to N_i (and for given T, the legal patent lifetime):

$$E(\pi_i) = \lambda N_i E(V) - N_i - F, \quad (6)$$

where

$$E(V) = \Pi_M \lambda \left(\sum_i N_i\right) \int_0^T e^{-\left(r+\lambda\sum_i N_i\right)\tau} d\tau + \frac{\Pi_M}{r}\left(1-e^{-rT}\right) e^{-\left(\lambda\sum_i N_i\right)T}. \quad (7)$$

In equation (6), λN_i is the Poisson arrival rate of innovation for firm i devoting to R&D an amount of resources (final output) equal to N_i, with $\lambda > 0$ being a parameter reflecting the productivity of the research technology. Apart from foregone consumption (N_i), there is also a fixed cost (F) that firms must pay in order to engage in R&D. At any point in time, firm i innovates with probability λN_i and, if successful, accrues the value of the patent, V. Equation 7 gives the expected value of a patent, where r denotes the real interest rate.

A firm that obtains a certain quality improvement keeps the private property of this innovation only during the legal life of patent (T). However, if the next innovation arrives before T elapses, then the actual lifetime of the patent will be limited by the duration between these two consecutive innovations. As a Poisson process is assumed for innovations, the duration between two consecutive innovations follows an exponential distribution with an arrival rate of $\left(\lambda \sum_i N_i\right)$. Accordingly, the expected value of a patent given in equation (7) is composed of two terms. The first term captures the probability of having the next innovation before T elapses and accruing the monopoly profit (Π_M) until when the new innovation shows up. The second term, instead, takes into account the probability that the next innovation arrives after T elapses so that the duration of the monopoly power accruing to the intermediate firm will be limited by T. Note that the knowledge covered by the patent will be public after T, leading to a competitive price in the intermediate sector till the next innovation occurs.

Concentrating on a symmetric equilibrium, we analyze the case where all R&D firms devote the same amount of resources (final output) to R&D in order to improve the quality of the existing capital good. The number of symmetric firms undertaking R&D (S) is bounded - as there exists a fixed cost (F) for engaging in R&D activity - and is assumed to be known by the government before determining the optimal (growth-maximizing) T.[9] Thus, assuming that $N_i = N$ for all i and denoting Π_M by Π, the first order condition for the maximization of $E(\pi)$ with respect to N (for given T) yields:

$$\frac{\Pi\lambda^2 NS}{r+\lambda NS}\left(1-e^{-(r+\lambda NS)T}\right)+\frac{\Pi\lambda\left(1-e^{-rT}\right)\cdot e^{-\lambda NST}}{r}+\frac{\Pi\lambda^2 Nr}{\left(r+\lambda NS\right)^2}\left(1-e^{-(r+\lambda NS)T}\right)+ \\ +\frac{\Pi\lambda^3 N^2 TS\cdot e^{-(r+\lambda NS)T}}{r+\lambda NS}-\frac{\Pi\lambda^2 NT\left(1-e^{-rT}\right)\cdot e^{-\lambda NST}}{r}=1 \tag{8}$$

4. Economic Growth

Using equations (2) and (4) and recognizing that in equilibrium $P_M = \frac{1}{\alpha}$, the level of aggregate output (Y) can be recast as:

$$Y = L^{1-\alpha}\left[q^M\left(\alpha^2 q^{\alpha M}\right)^{1/(1-\alpha)}L\right]^{\alpha} = L\alpha^{2\alpha/(1-\alpha)}Q$$

where $Q \equiv q^{\frac{\alpha}{1-\alpha}M}$ is an aggregate quality index relative to the available highest quality capital good. Hence, with L being constant, the growth rate of output is equal to the growth rate of Q. In turn, the proportionate change in Q due to a successful innovation is $\left(q^{\frac{\alpha}{1-\alpha}}-1\right)$

[9] In other words, we assume that the policy maker is a rational agent (in an absolute sense) and, as such, s/he is able to compute at each time the number of firms doing R&D.

and, under the symmetric equilibrium hypothesis, the flow probability of a success due to a Poisson process is λNS. Accordingly, the expected growth rate of the economy can be written as:[10]

$$g \equiv E\left[\frac{\dot{Y}}{Y}\right] = \lambda NS\left(q^{\frac{\alpha}{1-\alpha}} - 1\right). \tag{10}$$

In order for the aggregate economic growth rate (g) to be positive, the amount of resources devoted to R&D by each firm *(N)* should be positive, as the number of firms engaged in this activity is bounded from above. Checking the first order condition given in (8) when $S \to 0$ and $S \to \infty$ leads in turn to the restrictions $\lambda\Pi > r$ and $\lambda\Pi = 1$, respectively, for N to be strictly greater than zero.[11] $\lambda\Pi$ can be interpreted as the expected rate of return from investing one unit of foregone consumption in R&D activity. Together, the two restrictions we have just found say that this return should be bigger than the real interest rate in order for R&D firms to have sufficient incentives to invest in innovation.

4.1. Growth-Maximizing Patent Lifetime

In a symmetric equilibrium, and knowing the number of incumbent firms doing R&D (S), the government's problem of setting the legal patent lifetime that maximizes economic growth reduces to maximizing N with respect to T. Implicit differentiation of (8) yields the following condition for a growth-maximizing patent lifetime:

$$\frac{\partial N}{\partial T} = 0 \Rightarrow \tag{11}$$

$$(1-\lambda NT)(1+\lambda NS)\cdot e^{-(r+\lambda NS)T} + \frac{\lambda N(\lambda NST - S - 1)(1-e^{-rT})\cdot e^{-\lambda NST}}{r} + \frac{\lambda Nr\cdot e^{-(r+\lambda NS)T}}{r+\lambda NS} = 0$$

The formal solution to the government's problem can be obtained through solving the equations system (8) and (11) simultaneously. Unfortunately, it is impossible to conduct a complete algebraic analysis and obtain a closed-form solution. Therefore, we have proceeded with numerical simulations. These simulations show that, under the condition $r < \lambda\Pi < 1$, a finite growth-maximizing patent lifetime does exist and is unique.

5. Numerical Analysis

The aim of our numerical analysis is both to characterize the legal patent lifetime that maximizes economic growth and to see how it changes with respect to the level of

[10] Note that the growth rate of the economy does not depend on the quality level of the top generation capital good (M), but only on the increments of M due to the R&D investment. Hence, the government does not pay any attention to the intermediate sector market structure (whether it is monopolistic or competitive). This aspect of our model completely differs from the traditional, welfare maximizing patent lifetime models.

[11] The restriction $\lambda\Pi > r$ is obtained under the assumption that T is sufficiently large ($T \to \infty$).

competition in R&D, the real interest rate (r), the monopoly profit (Π) and the research technology productivity parameter (λ).

Figure 1 can be used to illustrate the legal patent lifetime, T^*, that will be set by a government in order to maximize economic growth and the corresponding amount of resources devoted to R&D by each firm, N^*, under the parameter values reported in Table 1:[12]

Table 1. **Parameter Values and corresponding T* and N***

$\lambda = 0.0005$	$\Pi = 500$	$r = 0.05$	$S = 100$	$T^* = 10.354$	$N^* = 1.575$

In Figure 1, according to the conditions given in (8) and (11), we have plotted two curves with a continuous and a dashed line, respectively.

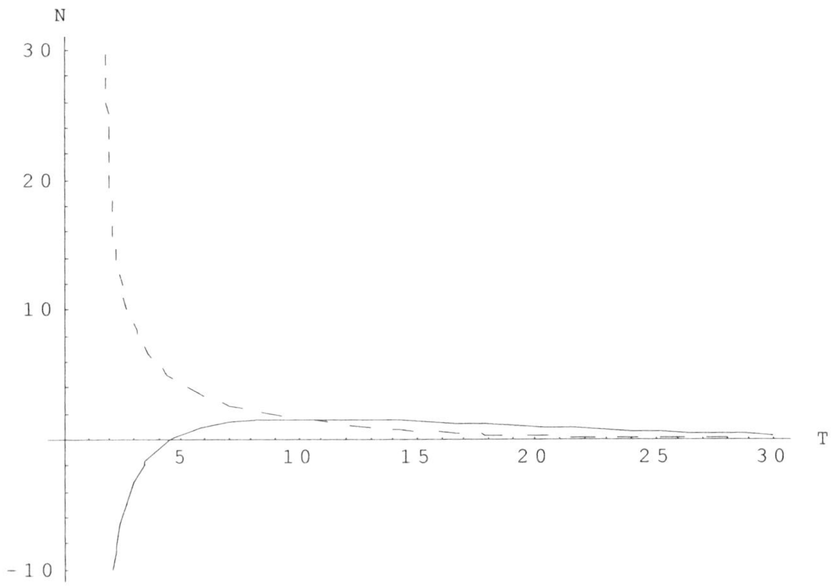

Figure 1. Characterizing the Growth-Maximizing Patent Lifetime

For a given T, the representative R&D firm determines the amount of resources devoted to innovation that maximizes its own instantaneous expected profit. This is reflected by the continuous line. On the other hand, given the profit-maximizing amount of resources that each firm will devote to R&D, the government chooses the patent lifetime that maximizes aggregate economic growth (the dashed line). The intersection of the two gives the solution to the government's problem (T^*). As it is clear from the figure, for admissible parameterizations, a finite growth maximizing patent lifetime exists and is unique.

[12] We have assigned the value of λ in the line of the empirical findings of Duguet and Kabla (1998). For the sake of expositional simplicity, we have assigned arbitrary values to Π, being aware of the fact that, indeed, it depends on the parameters α, q and M and on the size of the economy's labor force (L). Finally, we have set a usual value for the interest rate (r).

Our numerical simulations also suggest that an increase in the number of R&D firms (S), thus the level of competition in the innovation sector, leads to a decrease in the amount of resources devoted to research by each firm (see Table 2 below). The reason behind this result is simple. It is clear from the Table that the total R&D investment (NS) increases with S.[13] This implies a decline in the expected duration that would elapse between two consecutive innovations, which in turn reduces the incentive of each firm to engage in R&D. To compensate for this the government sets a higher legal patent lifetime in order to maximize N, hence economic growth. In Table 2, for given parameter values (λ, Π and r), we have reported T^* and the corresponding N^* values with respect to different numbers of firms doing research (S).

Table 2. **The impact of the level of competition in R&D (S)**

$\lambda = 0.0005$; $\Pi = 500$; $r = 0.05$	S = 1	S = 10	S = 100	S = 1000	$S = 10^6$
T^*	9.394	10.297	10.354	10.357	10.358
N^*	71.128	14.412	1.575	0.159	0.016

We also perform numerical exercises in order to analyze the effect of different r, λ and Π on T^*. The results are summarized in the following Tables 3, 4 and 5 below.

Table 3. **The impact of the real interest rate (r)**

$\lambda = 0.0005$; $\Pi = 500$; $S = 100$	$r = 0.05$	$r = 0.10$	$r = 0.15$	$r = 0.20$
T^*	10.354	10.921	11.983	14.238
N^*	1.575	1.053	0.608	0.247

Table 4. **The impact of the research technology productivity parameter (λ)**

$r = 0.05$; $\Pi = 500$; $S = 100$	$\lambda = 0.0002$	$\lambda = 0.0005$	$\lambda = 0.0010$	$\lambda = 0.0015$
T^*	29.175	10.354	4.669	2.709
N^*	0.761	1.575	2.320	3.455

A higher interest rate reduces the expected value for a patent, as firms would discount future profits more. As a consequence, due to a decrease in the expected net gain per unit of resources $(\lambda\Pi - r)$ devoted to R&D, the incentive to innovate reduces (N^*). Thus, the optimal legal patent lifetime maximizing economic growth (T^*) must increase. On the other hand, and following a similar line of reasoning, an increase in the expected net gain per unit of resources (due to an increase in either λ or Π) increases the incentive to innovate, so that a lower patent lifetime would suffice to reach the maximum level of economic growth.

[13] This result confirms what Loury (1979) has found concerning the relationship between the number of R&D firms, the total amount of R&D investment and the expected time to invention. However, he has not dealt with economic growth and thus growth-maximizing patent lifetime, which is the main concern of this chapter.

Table 5. **The impact of monopoly profit (Π)**

$\lambda = 0.0005$; $S = 100$; $r = 0.05$	$\Pi = 200$	$\Pi = 500$	$\Pi = 1000$	$\Pi = 1500$
T^*	29.175	10.354	4.669	2.709
N^*	0.305	1.575	4.640	10.365

6. Conclusion

Studying the possible economic effects of patent policy is at the core of modern research interests of economists. Textbook theory suggests that there are at least two fundamental ways through which patents may influence efficiency. On the one hand, patents generate welfare losses associated with the presence of temporary monopolies in the use of a new product/technology. This is the static, short-run, partial-equilibrium effect of patents. On the other, since Schumpeter (1942) the literature recognizes the need for patent protection to stimulate industrial R&D and economic growth. This is the dynamic, long-run, general-equilibrium effect of patents. Since patent policy may affect economic outcomes (*welfare* vs. *growth*) in conflicting ways, it is not that obvious how to strike a balance between static and dynamic objectives.

Until now, the patent design literature has, for the most part, confined itself to partial-equilibrium analysis. In this chapter we focused on one specific tool of patent policy (*patent lifetime*) and tried to ask the following question: What is the optimal patent length a government would set in order to maximize economic growth within an endogenous growth model with vertical differentiation and stochastic R&D activity? The originality of the analysis derives from the fact that, although the endogenous-growth literature has recognized the need for patent protection to stimulate R&D effort and growth, thus far attention has been limited to one simple policy, often referred to as *"infinitely-lived patents"*.

To answer the question above we considered an economy with three vertically integrated sectors. A homogeneous consumption good is obtained by combining labor and an intermediate input (available in different qualities) through a Cobb-Douglas technology. The diverse quality grades of the existing type of intermediate input are perfect substitutes (the discovery of a higher grade implies that the lower grades are driven out of the intermediate sector). Firms that operate in the R&D sector race to produce innovations in terms of quality improvements in the existing capital good product line. In deciding how many resources to devote to research, R&D firms consider two variables: the size of the profit flow coming from this activity and its likely duration. This duration is random, as it depends on the uncertain outcomes from the research efforts of competitors. A policy maker knows how many R&D firms are racing at each date for a new innovation and in a symmetric equilibrium (where R&D firms are all identical to each other) sets the patent lifetime in order to maximize the amount of resources devoted to innovation by the representative R&D firm and, then, the aggregate economic growth rate.

Through numerical simulations we show that, under plausible parameterizations, a finite growth maximizing patent lifetime does exist and is unique. We also show that the optimal, growth-maximizing patent lifetime increases with the level of competition in the R&D sector and the real interest rate, whereas it decreases with the monopoly profit accruing to the monopolistic capital good producer and the productivity parameter of research technology.

In the light of these results, two important questions still remain open in the future research agenda. First, along with patent lifetime, the simultaneous impact of other policy instruments (such as patent breadth) on R&D investment and economic growth would be analyzed. Secondly, the fact that the same economic concept - monopoly - can have positive or adverse economic consequences (on the private incentives to innovate and on social welfare, respectively) makes it of paramount importance to define accurately the boundaries between intellectual property rightsand antitrust laws.

REFERENCES

Aghion, P., & Howitt, P. (1992). A model of growth through creative destruction. *Econometrica*, **60**(2), 323-51.

Dasgupta, P., & Stiglitz, J. (1980). Industrial structure and the nature of innovative activity. *Economic Journal*, **90**(358), 266-93.

Duguet, E., & Kabla, I. (1998). Appropriation strategy and the motivations to use the patent system: An econometric analysis at the firm level in French manufacturing. *Annales d'Economie et de Statistique*, **0**(49-50), 289-327.

Eaton, J., Kortum, S. & Lerner, J. (2004). International patenting and the European patent office: A quantitative assessment". In *OECD (Ed.), Patents, Innovation and Economic Performance* (pp. 27-52), Paris: OECD Publications.

Grossman, G.M., & Helpman, E. (1991). *Innovation and growth in the global economy*. Cambridge, MA: MIT Press.

Helpman, E. (1993). Innovation, Imitation, and Intellectual Property Rights. *Econometrica*, **61**(6), 1247-80.

Horowitz, A.W., & Lai, E.L.C. (1996). Patent Length and the Rate of Innovation. *International Economic Review*, **37**(4), 785-801.

Judd, K.L. (1985). On the performance of patents. *Econometrica*, **53**(3), 567-85.

Keely, L.C. (2001). Using patents in growth models. *Economics of Innovation and New Technology*, **10**(6), 449-92.

Loury, G.C. (1979). Market Structure and Innovation. *Quarterly Journal of Economics*, **93**(3), 395-410.

Martínez, C., & Guellec, D. (2004). Overview of recent changes and comparison of patent regimes in the United States, Japan and Europe. In *OECD (Ed.), Patents, Innovation and Economic Performance* (pp. 127-62), Paris: OECD Publications.

Michel, P., & Nyssen, J. (1998). On knowledge diffusion, patents lifetime and innovation-based endogenous growth. *Annales d'Economie et de Statistique*, **0**(49-50), 77-103.

Nordhaus, W.D. (1969). An economic theory of technological change. *American Economic Review*, **59**(2), 18-28.

O'Donoghue, T., & Zweimüller, J. (2004). Patents in a model of endogenous growth. *Journal of Economic Growth*, **9**(1), 81-123.

OECD (2002). *OECD Information Technology Outlook - Highlights*. Paris: OECD Publications.

OECD (2004a). *Patents, innovation and economic performance*. Paris: OECD Publications.

OECD (2004b). *Patents and innovation: Trends and policy challenges*. Paris: OECD Publications.

Romer, P.M. (1990). Endogenous Technological Change. *Journal of Political Economy*, **98**(5), S71-S102.

Schumpeter, J. (1942). *Capitalism, socialism and democracy*. London: Allen and Unwin.

Segerstrom, P.S., Anant, T.C.A., & Dinopoulos, E. (1990). A Schumpeterian model of the product life cycle. *American Economic Review*, **80**(5), 1077-91.

In: Economics of Technological Change
Editor: Thomas L. Beane, pp. 39-67
ISBN: 978-1-60021-237-6
© 2007 Nova Science Publishers, Inc.

Chapter 3

THE EFFECTS OF INNOVATION ON PERFORMANCE OF KOREAN FIRMS

Almas Heshmati[*]
University of Kurdistan – Hawler and Techno-Economics & Policy Program, College of Engineering, Seoul National University, San 56-1, Shinlim-dong, Kwanak-gu Seoul 151-742, Korea
Yee-Kyoung Kim[**] *and Hyesung Kim*[***]
University of Kurdistan – Hawler and Techno-Economics & Policy Program, School of Economics, Seoul National University, San 56-1, Shinlim-dong, Kwanak-gu Seoul 151-742, Korea

ABSTRACT

This study empirically examines the relationship between knowledge capital and performance heterogeneity at the firm level. The model is based on a knowledge production function comprising of four interdependent equations linking innovativeness to innovation input, innovation output and productivity. The empirical part is based on Korean firm level innovation data. The model is estimated using advanced econometric methods. We investigate whether innovation is a significant and contributing determinant of performance heterogeneity among firms. In examining the relationship between innovation and productivity we correct for selectivity and simultaneity biases. The results show that there is a two-way causal relationship between knowledge capital and labor productivity. Firm-specific effects positively contribute to innovation output but they are negatively related to productivity. Industry heterogeneity does not affect innovation output or productivity.

JEL Classification Numbers: C33; E22; L60; O32

Keywords: Innovation Input; Innovation Output; Productivity; Korea

[*] E-mail address: heshmati@snu.ac.kr, Tel: +82-2-880-9141 Fax: +82-2-880-8389
[**] E-mail address: icejean@freechal.com, Tel: +82-16-595-6148
[***] E-mail address: hskim001@snu.ac.kr, Tel: +82-2-881-5663

1. INTRODUCTION

Over the last few decades, Korea has successfully achieved rapid economic growth. Right after the Korean War, Korea was merely one of many poor countries whose per capita GDP was $300 as of 1955. 40 years later in 1995, Korea's per capita GDP exceeded $10,000 and as of 2004, Korea's per capita GDP was $14,100 (OECD, 2006). Such Rapid growth has been the source of admiration. There are many factors which believed to be important to such rapid economic growth such as cheap and disciplined labor and Korean government's industrial policy. Korea's growth oriented industrial policy, indeed, took advantage of cheap labor in the earlier period of growth.

Table 1. **R&D Investment**
Unit: 10 billion Won

	R&D Investment			**Contribution**		**R&D Investment Growth***		
	Gov't	**Private**	**Total**	**Gov't**	**Private**	**Gov't**	**Private**	**Total**
1980	1.80	1.03	2.83	63.70	36.30	-	-	-
1981	2.02	1.65	3.67	55.00	45.00	-4.40	37.60	10.90
1982	2.64	2.69	5.33	49.60	50.40	23.70	53.50	37.10
1983	2.31	4.51	6.82	33.90	66.10	-17.60	58.10	20.60
1984	2.52	6.56	9.07	27.70	72.30	2.70	37.20	25.50
1985	3.07	9.30	12.37	24.80	75.20	15.80	34.70	29.50
1986	3.74	12.33	16.07	23.30	76.70	16.00	26.00	23.50
1987	4.90	14.95	19.85	24.70	75.30	24.30	15.10	17.20
1988	5.23	19.31	24.54	21.30	78.70	-1.50	19.20	14.10
1989	5.75	22.42	28.17	20.40	79.60	3.10	8.80	7.60
1990	6.51	26.99	33.50	19.40	80.60	3.00	9.50	8.20
1991	8.16	33.43	41.58	19.60	80.40	12.50	11.20	11.50
1992	8.79	41.11	49.89	17.60	82.40	-0.40	13.70	10.90
1993	10.39	51.14	61.53	16.90	83.10	11.00	16.70	15.70
1994	12.60	66.35	78.95	16.00	84.00	12.50	20.30	19.00
1995	17.81	76.60	94.41	18.90	81.10	31.90	7.70	11.60
1996	23.98	84.80	108.78	22.00	78.00	29.50	6.50	10.90
1997	28.51	93.35	121.86	23.40	76.60	15.90	7.30	9.20
1998	30.52	82.85	113.37	26.90	73.10	2.50	-15.00	-10.90
1999	32.03	87.19	119.22	26.90	73.10	6.80	7.10	7.00
2000	34.52	103.97	138.49	24.90	75.10	9.40	21.00	17.90
2001	41.87	119.23	161.11	26.00	74.00	18.10	11.70	13.30
1981-1989	-	-	-	26.40	73.60	6.90	32.30	20.70
1990-2001	-	-	-	21.50	78.50	12.70	9.80	10.40
1981-2001	-	-	-	25.70	74.30	10.20	19.40	14.80

Note: *Real growth rate is calculated after being deflated by using GDP deflator

Source: Survey on Research and Development Activities for Science and Technology, Ministry of Science and Technology, 2002

However, what should be noted is that in the progress of development of its industrial development strategy, Korea has put great emphasis on innovation and to build up the country's scientific and technological capacities, which were in fact so crucial in realizing vision of Korean growth and development. Branscomb and Choi (1996) assess the assets, institutions, and policies needed to master an innovation-based development strategy. Hence, the essence of Korea's remarkable economic growth lies in the increase in the stock of useful knowledge and the continuous innovation effort. In this new millennium, the importance of knowledge and technology has become even greater than ever and as such, Korea is committed to technology-based innovation as the primary source of its economic growth and social development. Such efforts and commitment have been made both in the public and private sectors. Table 1 summarizes R&D investment both by government and in private sector from 1980 to 2001.

R&D investment is merely seen as one of the main indicators of technological or innovation capability of a country. If innovation capability of a country comes from innovation capability of industries and firms, government's vision and industrial policy on innovation can influence to a great extent the innovation capability of a country through influencing innovation capability of firms and industry. In this sense, Korea's science and technology based economic growth can be attributed greatly to the government's industrial policy aiming at building up the technological capability.

There has been a large volume of work in examining the link between the innovation and productivity, and several studies find evidence of a positive effect of innovation on productivity. One of the most significant short-comings of common approaches is that it is the innovation output, not innovation input, which causes productivity to increase and this issue is raised and resolved by Crepon, Duget and Mairesse (1998), hereafter labeled as CDM, in their significant contribution to the literature.

Following the CDM approach and applying Community Innovation Survey data, Lööf and Heshmati (2002) examined the heterogeneous impacts of knowledge capital on firm level productivity growth. With these two papers as our benchmark in terms of methodology, our study takes into account that innovation output, not innovation input, in fact influences productivity and attempts to examine the link between innovation output and productivity by using the Korean Data. The community innovation survey data based on Korean manufacturing firms, namely Korean information System (KIS), is an extensive dataset covering 3,775 firms. Despite the usefulness of data, since the dataset has been constructed quite recently, it has not widely been applied in empirical studies. As such, this paper can be distinguished from other work on the innovation and productivity relationship using Korean firms in the sense that we examine innovation input-output-productivity chain and that we utilize Korean Innovation Survey data.

This study empirically examines knowledge capital and productivity at the firm level based on Korean firm level data. By applying the advanced econometric method to Korean Community Innovation Survey (CIS) data, we investigate whether innovation is a significant and contributing determinant of productivity heterogeneity across firms. In examining the relationship between innovation and productivity we successfully correct for selectivity and simultaneity biases.

The rest of the study is organized as follows. In Section 2 we provide some description about Korean manufacturing industries. In Section 3, we provide a review of the literature. The theoretical approach of this paper is presented in Section 4. Section 5 provides the description of our data and variables. Section 6 presents the empirical model and estimation method. Empirical results are given in Section 7, followed by the conclusion and discussion of the paper in Section 8.

2. THE KOREAN MANUFACTURING INDUSTRIES

Korean industrial policies historically pursued by Korean governments can be characterized as export-oriented policies and is sometimes subject to the criticism for the strong concentration on enterprise grouping (Chaebols) due to focus on capital-intensive and high-tech intensive technologies contributing to weakening small and medium enterprises (SME) and the technological fragility. Nevertheless, despite the criticism, it is no doubt that such industrial policies are a main contributor to the rapid convergence of Korean industrial structure to the industrial structure of the advanced countries.

The comparative analysis of changes of industrial structure and productivity of Korea and G7 developed countries[1] during the last 30 year (1970-2000) was conducted by Korea Development Institute (KDI). The structural composition of value added and employment was calculated by using the OECD STAND database in five major industries such as textiles and textile products, chemical products, machinery, electrical and electronic equipment, and transport equipment. The overall empirical results show that Korean industrial structure converges to those of G7 countries along with narrowing productivity gap. Beyond these expected observations, the international comparison shows that the speed of structural change in Korea has been much faster than other developed countries. It indicates the dynamism of Korean economic development and presents a challenge facing current Korean economy by bringing about growing sectoral productivity divergence as called economic polarization.

As mentioned above the first and foremost characteristic of Korean industrial trend in comparison with other developed countries is the dynamism of structural change. According to the employment structure, the most important index of structural transformation, Korea has experienced both industrialization and de-industrialization recently. While the employment shares of manufacturing sectors in other developed countries have steadily declined since 1970, Korea experienced a rising manufacturing employment until the late 1980s while it declined rapidly afterwards. As a result in 2000, the employment composition of Korea looks similar to that of G7 countries. Despite declining employment, however, Korean manufacturing sector has kept above 30% of total value added in the 1990s, while the manufacturing share in total valued added in G7 countries declined toward about 20% on the average. The decline of manufacturing employment in the 1990s led to the increase in service sector employment, particularly in the traditional low paying jobs. It brings about widening sector difference in labor productivity, so called polarization of sectoral growth and income. For details on the above comparisons see Figures 1 to 3.

The decline in manufacturing employment resulted from the decline of the light industries such as the textiles after the 1980s. Although heavy and chemical industries continued to provide new jobs, the sharp employment reduction in light industries decreased the total employment of manufacturing sector in the 1990s. It implied the rapid transition of Korean manufacturing sector toward more capital and technology intensity. Among heavy and chemical industries, electrical and transport equipment including automobiles show relatively higher performance than machinery. In terms of labor productivity measured as per worker valued added, the electrical equipment including electronic products achieves more

[1] The G7 countries are USA, UK, France, Germany, Italy, Japan and Canada.

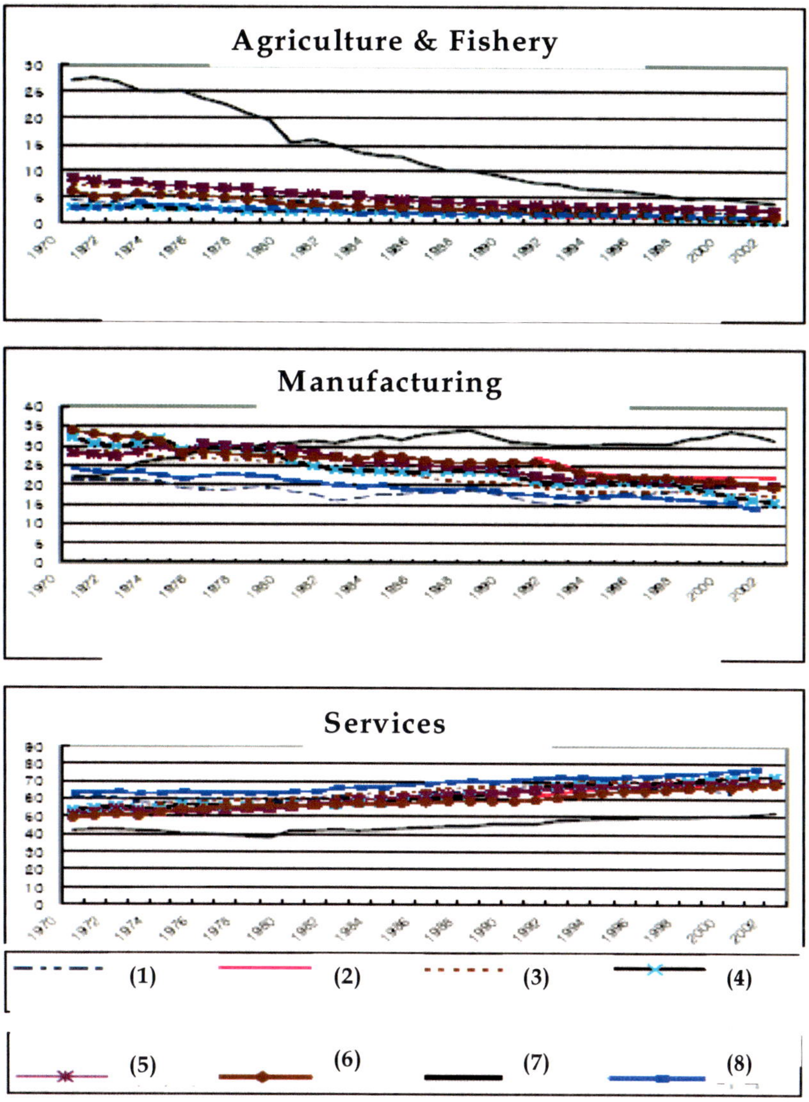

(1) Canada (2) Germany (3) France (4)Great Britain
(5) Italy (6) Japan (7) Korea (8) USA

Figure 1. The structural composition of vale added(1970-2000)

than 70% of the U.S. level by surpassing some G7 countries. However, the labor productivity of machinery stays low at the 25% of the U.S. level. The labor productivity of textile industry continued to stay lower than 10% until 2000.

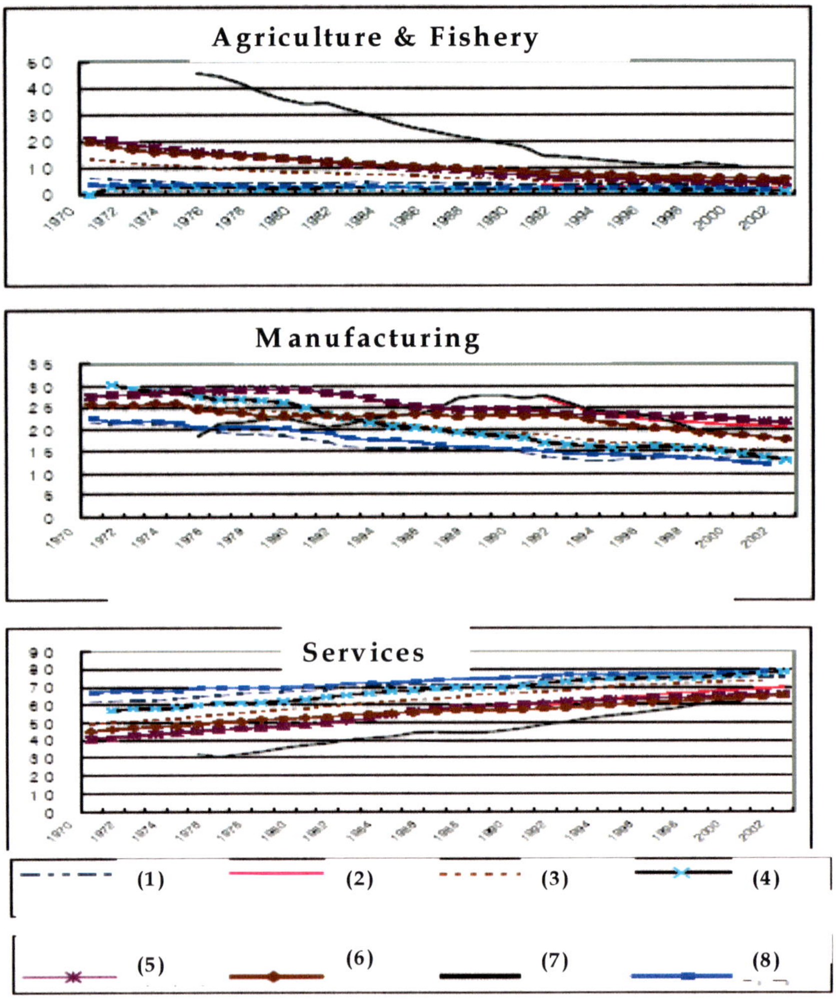

(1) Canada (2) Germany (3) France (4)Great Britain
(5) Italy (6) Japan (7) Korea (8) USA

Figure 2. The structural composition of employment (1970-2000)

3. Review of the Literature

3.1. Korean Innovation Research

Until recently, very little studies on innovation using the Community Innovation Survey have come out since Korean CIS has been conducted only for the past few years. Thus, this area is fairly new in the Korean literature. After the Korean CIS data is released to the public, innovation studies based on this database have attracted many researchers and many research projects are currently in progress. On the 21st October 2004, an innovation workshop took

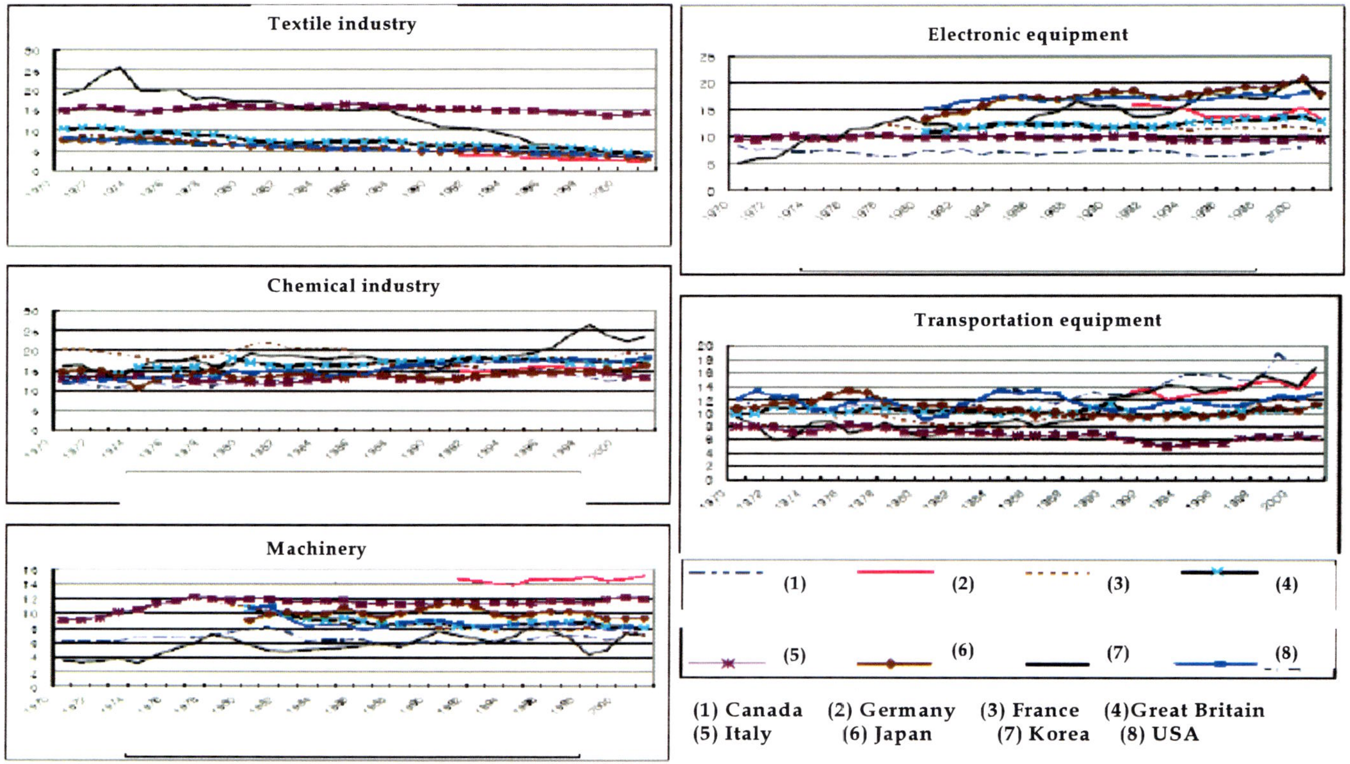

Figure 3. The structural composition of value added in selected manufacturing industries. (1970-2000)

place for the recent papers that is based on CIS data at Science and Technology Policy Institute (STEPI hereafter), in Seoul Korea. All the papers presented at the workshop were based on KIS (Korean Innovation Survey) 2002 released by STEPI. A summary of some studies follows.

Sung (2004) analyzes the determinants of firm's innovative activity, focusing especially on the role of inter-firm networks and technological opportunity in performing innovative activities. To measure innovative activity, he adopted three different variables in turn as proxies for innovative activity, namely the product innovation, product improvement, and process innovation. The explanatory variables such as firm size, market concentration ratio, lagged profitability, foreign ownership, export ratio, firm's age, and formal R&D activity are also considered. With data from 1,124 firms for the two years (2000-2001) Sung estimated the logistic regression model, and found that the determinants of firm's innovative activities differ according to both innovation type and technological opportunity and also that the network effects also differ according to both innovation type and technological opportunity. Furthermore, he found that high-technological-opportunity firms are more innovative than low-technological-opportunity firms not in product improvement or process innovation, but in product innovation.

Seo (2004) investigates the relative importance of various sources of technological innovations and their effects by conducting factor analysis. In this study, the selected factors of the sources of technological innovations based on the factor analysis are university/research institutes, firms' internal sources, general information, vertical/competition relationship sources, and horizontal relationship sources. Regression analysis is conducted in order to examine the effects of each source on technological innovative activities, and finds that firms' internal sources have the greatest impact on innovative activities in terms of both innovation input and output among the variables whose impacts are statistically significant. For the innovation input, as the horizontal sources are abundant, the less incentives for innovation activities for firms. For the innovation output, the more abundant sources from university/research institutes, the more innovation, while the more access to general information media and more sources from vertical/competition relationship, the less the innovation activities. The firm size and the stage of technological innovations are found to cause the heterogeneity in the effects on technological innovation.

In Lee et al. (2004), variance decomposition method is used to see whether firm or industry effects have a greater between industry effect and firm effect on R&D intensity. The finding of the study suggests that the firm effect is greater than industry effect on R&D intensity, and the industry effect was in fact hardly influential except in non-high tech industries. Among the venture firms and in high-tech industries, industry and firm effects do not hold much explanatory power. Rather, much variation is captured in the residual effect. Also, regarding the difference in effects among different R&D types, it is found that firm effect is in general greater than the industry effect but those who have high intensity in the searching type R&D are found to exhibit greater industrial effect compared to those who have lower intensity.

Seo and Lee (2004) tried to find the hampering factors of firms' innovative activities also using the KIS (Korea Innovation Survey) 2002 data. Employing the method of factor analysis, they identified five hampering factors of innovative activities namely, organizational hampering factor, innovative hampering factor, economic hampering factor, supply-chain hampering factor, and market-environment hampering factor. Using regression analysis which includes the five hampering factors, industry characteristics variables (market concentration, technological opportunity, and appropriability), and other control variables, they found that: (1) all the five hampering factors are negatively related with innovative activities, (2) technological opportunity and appropriability conditions are positively related with innovative activities, (3) there is an inverted-U relationship between market concentration and innovative activities, and (4) firm size has no significant influence on firms' innovative activities.

It is worth to mention that, although the studies review above use the same source of innovation data, current study in no way overlap with those described above in its objective or methodologies employed.

3.2. Studies of Firm Heterogeneity and Productivity

In a large volume of empirical literature it has been found that the significant heterogeneity are present among firms and shown that much of the heterogeneity seems quite persistent over time. The evidence of persistency of leading firms is found in Klette and Griliches (1998), Roller and Sinclair-Desgagne (1996) and Lööf and Heshmati (2002). The tendency that the amount of productivity dispersion is considerable and persistent in nature implies that the higher the today's productivity of firms, the higher the tomorrow's productivity (Bartelsman and Doms, 2000). Moreover, according to the findings in Bartelsman and Doms (2000), heterogeneity in labor productivity and productivity growth is associated with similar heterogeneity in total factor productivity (TFP).

3.3. Sources of Productivity Growth

Analysis of productivity has been attracting enormously the researchers who conduct studies at firm level for many reasons: First of all, productivity is a key factor to determine firms' prosperity and the survival in the market. That is, firms with higher output growth has been found to be more productive and also less probability of exit. Highly productive plants have been found to have higher output growth and are less likely to exit. Foster, Haltiwanger and Krizan (1998) has shown that relative productivity between plants can be explained by wage difference, export success and technology utilization.

Regarding the sources of productivity, Griliches (1994) looks at TFP growth rate at the aggregate level in which 1.5–2.0 percent as the approximate growth rate per year is assumed, and finds that at least half of this amount is likely to be associated with growth in the quality of the labor force, economies of scale and various reallocations of capital between assets and industries. One should note that the remaining half can be attributed to advances in knowledge, commercialized innovations. There have been significant volume of literature which have examined the impact and its extent of research and development (R&D) capital on differences in productivity among firms. For a survey of econometric studies of R&D activities at the firm level, see Mairesse and Sasseneou (1991), and the discussion of difficulties associated with such approach.

3.4. Innovation Input and Innovation Output

A growing volume of literature has addressed itself to analysis of factors influencing innovative activities of firms over the past 2 decades. For the details in a recent survey, refer to Cohen and Levein (1989). Much of this literature focuses merely on examination of the determinants of innovative activity of R&D performing firms. However, not all firms are engaged in R&D activity; Rather, it is restricted to small proportion of firms. This feature is more common in developing countries.

Studies restricted to samples of firms engaged in formal R&D are, therefore, prone to a selectivity bias. Crepon, Duguet and Mairesse (1998) and Lööf and Heshmati (2002) have taken into account the selectivity bias in their studies: In both of the studies where R&D investment is defined as innovation input while Patent count and innovation share of sales as innovation output, an innovation equation and a productivity equation are treated as a system and innovation input is endogenous in the innovation output equation and innovation output is endogenous in the productivity equation. Since there is a major issue of simultaneity which is likely to interact with selectivity, they take into account both sources of biases. CDM (1998) shows that the conventional OLS regressions gives a negligible estimated elasticity of productivity with respect to innovation output and find that measurement errors should be the cause. To deal with error-in-variable problems, instrument variables are introduced, and the innovation output elasticity are found to increase significantly.

4. The Theoretical Model

Crépon, Duget and Mairesse (1998), henceforth CDM, tried to explain different results obtained in estimating the relationship between innovation and productivity growth. In estimating the relationship between innovation and productivity/productivity growth, the results in terms of estimates of R&D elasticity in the productivity/productivity growth equation have significantly varied. Mairesse and Sassenou (1991) in their survey of econometric studies of R&D and productivity at the firm level well document the issues of the presence of heterogeneity in estimation results and provide number of factors that cause such variation.

In the CDM article, the production function analysis was used but with some new features. First, in their model it is explicitly taken into account that it is not innovation input, but rather innovation output that increases productivity. CDM model includes following three relationships: (i) the innovation input linked to its determinants, (ii) the knowledge production function relating innovation input to innovation output, and (iii) the productivity equation linking innovation output to productivity growth. Second, CDM article uses new data on innovation output and variables indicating demand conditions, technological opportunities, concentration, diversification, and imitations. Such information that has been included in addition to typical firm information namely information on firm's current account, balance sheets, and R&D expenditures, takes into account the heterogeneity of firm behavior. Third, in estimating their model, CDM uses econometric methods that can address some of the problems present in models analyzing innovation data, namely selectivity and simultaneity biases.

The innovation studies that deal with R&D-Productivity link are often restricted to samples of firms that are engaged in formal R&D activities. However, considering the empirical evidence that only a minority of firms is involved in formal R&D activities, such restriction may result in selectivity bias. Also, in a model like CDM, which includes both innovation and productivity equations, and where innovation input is endogenous in the innovation output equation and innovation output is endogenous in the productivity equation, presence of simultaneity is an important issue when estimating the model. CDM finds that simultaneity tends to interact with selectivity and that both sources of biases, therefore, must be taken into account.

According to CDM article, the often-used OLS regressions give a negligible estimated elasticity of productivity with respect to innovation output, and measurement errors should be important in such case. In dealing with error-in-variable problems when instrumental variable is used, the innovation output elasticity increases from about 0 to about 0.3–0.4. We now present the original CDM model with its three main equations.

The econometric model of investment in the CDM model relies on two different equations. Not all firms invest in innovation, so that specific econometric methods need to be used. In a first step, firms decide whether to engage in research or not. This decision is modeled as a probit model. In the second stage, firms decide how much to invest in research conditional that thcy have already decided to make an investment. Let g_i^* be the unobserved decision variable whether to invest in R&D or not, and k_i^* be the unobserved level of investment of firm i. g_i and k_i are their corresponding observable counterparts.

The two-equation investment model as a function of observed vectors of characteristics x_i^0 and x_i^1 is written as:

$$g_i^* = x_i^0 \beta_o + u_i^0, \quad g_i = 1 \text{ if } g_i^* > 0, \text{ else } g_i = 0, \tag{1}$$

and

$$k_i^* \mid g_i^* > 0 = x_i^1 \beta_1 + u_i^1, \quad k_i = k_i^* \text{ if } k_i^* > 0, \text{ else } k_i = 0, \tag{2}$$

The β_0 and β_1 are vectors of unknown parameters to be estimated reflecting the impact of certain factors on the probability of engagement in investment and the actual level of R&D investment. The u_i^0 and u_i^1 are random error terms with mean zero, constant variances and not correlated with the explanatory variables. However, the two error terms are correlated with each other. The correlation coefficient suggests a selection based on the unobservable characteristics of firms with a high potential to be involved in R&D activities. The model can be estimated in a two-step sample selection procedure (probit in the first step and least squares in the second step). Hence, one of the explanatory variables in x_i^1 (equation 2) is Mill's ratio, calculated using the probit parameter estimates to correct for possible selection bias. Alternatively the R&D model in equations (1) and (2) can be estimated in one single step in a generalized tobit model using the maximum likelihood estimation method.

The CDM model outlined in above is based on the Griliches (1994) who describes the transformation process that goes from innovation input to economically valuable innovation output as the "knowledge production function". In the CDM model this equation is given by:

$$t_i = \alpha_k k_i^* + x_i^2 \beta_2 + u_i^2, \tag{3}$$

where t_i is the observed innovation output (sales), and α_k is the elasticity of innovation output with respect to innovation input. x_i^2 is a vector of explanatory variables, and u_i^2 is the random error term with mean zero and constant variance and not correlated with the x_i^2 variables. The last equation in the model describing the relationship between innovation output and productivity growth is labeled as the productivity equation expressed as:

$$\dot{q}_i = \alpha_t t_i + x_i^3 \beta_3 + u_i^3 \tag{4}$$

where $\dot{q}_i$ is the rate of productivity growth, α_t is the elasticity of productivity with respect to innovation output, and x_i^3 is a vector of other common explanatory variables in a Cobb–Douglas production function, including labor and capital intensity, skill composition and firm size. The error terms in the model reflect in part unobserved variables and in part firm-specific effects. CDM follow the generalized tobit model assumption of joint normality of the disturbances in the first two equations (probit and tobit models).

The model allows for arbitrary correlation among the four disturbances u_i^0, u_i^1, u_i^2 and u_i^3. The error terms are assumed to be independent of the exogenous variables in the different equations. In the second part, the innovation output and productivity equations are estimated as a system of equations. The procedure thereby accounts for both sources of bias, namely selection and simultaneity biases.

5. Data and Variables

5.1. KIS Data Description

This empirical analysis is based on the data collected by STEPI (2002) in their 2002 Community Innovation Survey for manufacturing firms. The KIS (Korea Innovation Survey) dataset which collects data about innovation activity at the firm level in Korea covers more than 3,775 firms and 21 industries. For the estimation, only 3331 observations were utilized after the firms whose productivity and employment in 2001 was 0 or missing due to the use of growth rate were removed. After then, several other missing units were replaced by 0.

In this study, we define innovations as products (goods and services) and processes that is: (i) new or substantially improved to the market and (ii) new or substantially improved only to the firm. The effects of both these classifications of innovation can be analyzed jointly or separately. We then define an innovative firm according to the following conditions: a firm is innovative if its innovation investment is positive, and if it also has positive innovative sales. Here by innovative sales, we refer to the sales revenue of a firm that is attributed to products introduced on the commercial market in the three most recent years.

This condition resulted in a sample of 1019(30.55%) innovative firms. The firms whose innovation investment and innovative sales both were 0 consist of 48.73% of the sample, i.e. 1625 firms. The firms with no innovation investment but with positive innovative sales resulted in a sample of 376 consisting of 9.45%, while the firms that have made positive innovation investment but not yet have positive innovative sales are 376 firms consisting of 11.27% of the total sample. The later 2 percentages are due to the fact that it takes time for an innovation investment to generate innovation sales, and also not all investments in innovation succeed in producing new products and processes.

Innovation activity in Community Innovation Survey is restricted to the activities that are introduced in the market and thus results in an increase in sales or reduction in costs. If the proportion of innovation active firms is high among the entire firms of a country, the capability of technological innovation in that country, especially in the particular sector, can be assessed to be high. Also the firms whose productivity per unit of labor was 0 resulted in a sample of 1940 (58.17%).

For the joint sample of innovative and non-innovative firms and separately for only innovative firms, summaries of the data are presented in Table 3. The summary statistics Table 3 shows that mean values of Korean innovative firms are higher than total firms in Korea in most respects, as was expected. Especially, the average innovation input (innovinput) of innovative firms is about 2.5 times higher than that of average firm. The average innovation output (innovoutput) of innovative firms is about twice higher than that of average firm. Innovative firms achieved more sales about 1.6 times compared to a average firm. The size of innovative firms with average number of 192 employees is larger than that of total firms with average number of 120 employees. The average number of patents of innovative firms is 7.5 which are about twice higher then that of average firms with 3.5. The number of researchers employed by average firm is 11, which is half of that of an innovative firm with 24 researchers on average. In contrast, the labor productivity of innovative firms is 235 and it is lower than that of an average firm which is 224.

Since the correlation matrix in Table 4 shows that productivity is increasing with the more innovative output. And innovation output increases by the more innovation investment. Patents are negatively correlated with the productivity. Researchers are highly correlated with productivity. Multicollinearity was not found to be a distinctive problem among variables except for researcher and export.

A variety form of variables encompasses continuous variables and discrete variables which are ordered rank variables and dummy variables. Most of KIS data are ordered ranking variables which range from 0 to 6. The two samples, namely total sample and innovation sample, are further divided into three groups based on the firms' evaluation of their degree of importance, strongly, quite or moderately, when the innovation strategy and sources of knowledge are considered. Due to the nature of the model, the available information is divided into four panels, namely: firm innovation strategy, sources of knowledge for innovation, co-operation on innovation and factors hampering innovation. For more details please refer to Appendix A.

Table 2. Labor productivity

		1975	1980	1985	1990	1995	2000
All industries	Canada	33.2	34.8	37.5	38	40.7	44.8
	Germany	.	.	.	.	52.5	56.5
	France	.	42.3	47.6	53.2	56.1	59.4
	Great Britain	26.8	28.6	33	35.3	40.5	44.7
	Italy	30.6	35.2	37.9	41.9	45.9	48.1
	Japan	42	50.5	57.6	69.3	72.6	78.3
	USA	47.1	48.5	52.5	55.3	58.8	64.9
	Korea	7	8.4	11.2	14.6	18.5	22.6
Manufacturing industries	Canada	32.2	36.4	44.7	46.6	56	67.1
	Germany	.	.	.	.	52.6	57.6
	France	.	38.5	43.7	50.7	61.4	73.3
	Great Britain	27.4	28.1	36.5	44.4	53.6	59.3
	Italy	23.9	28.3	32.3	37.2	43.5	45.9
	Japan	36.6	48.3	58.2	71.2	78.3	95.4
	USA	38.7	42.8	53.5	60.7	73	90.7
	Korea	6.6	9.1	12.2	15.7	24.2	39.9

Source: KDI Comprehensive Study on Industrial Competitiveness of Korea (II) (2004)

Note: Unit: Korean Million won. Reference year is 1995. Germany is unified Germany including East Germany.

Table 3. Summary Statistics of data : Numeric variables

	Total firms (N=3335)				Innovative firms (N=1391)			
Variable	**Mean**	**Std Dev**	**Min.**	**Max.**	**Mean**	**Std Dev**	**Min.**	**Max.**
employment	129.473	589.466	1	25950	191.574	815.929	3	25950
productivity	235.024	1251.499	0.19	58599	224.336	536.692	0.86	15340.61
Total Sales	39244.882	277444.817	5	9999000	65048.023	390697.031	44	9999000
Export	11550.291	111339.71	0	4914301	21458.999	167109.229	0	4914301
Profit	2686.581	15622.279	0	532311	4458.934	21867.951	0	532311
Equity	10093.838	109586.695	-151	4001200	16002.861	124998.788	0	3003200
researchers	11.355	181.893	0	10210	24.239	280.708	0	10210
innovinput	484.26	5151.905	0	260799	1161.037	7929.476	2	260799
innovoutput	13830.384	190201.286	0	7404230	28792.977	292819.285	0	7404230
Patents	3.573	26.058	0	780	7.508	39.464	0	780

5.2. Variables and their Definitions

• ***Innovation investments*:**

Innovation investment is broken down into three different categories. In the Korean Community Innovation Survey it is defined as R&D investment including wages and labor costs for R&D personals. This cost can further be divided into the internally financed R&D spent for in-house R&D, external R&D costs funded internally, and internal R&D spending supported by external sources. We combine these three costs to generate the innovation investment, and based on the definition we found that nearly 35 percent of all firms made an investment in some innovative activities, according to the KIS definition, in 2001.

Table 4. Correlation matrix : innovative sample

	Innov-input	Innov-output	Product-ivity	Employ-ment	Research-ers	export	Patents
innovinput	1	0.1768	0.0069	0.2090	0.0500	0.1601	0.2778
		0.0001	0.6867	0.0001	0.0038	0.0001	0.0001
innovoutput	0.1768	1	0.0483	0.7080	0.0792	0.8435	0.3292
	0.0001		0.0052	0.0001	0.0001	0.0001	0.0001
productivity	0.0069	0.0483	1	0.0119	0.0013	0.0624	-0.0022
	0.6867	0.0052		0.4901	0.9364	0.0003	0.8988
employment	0.2090	0.7080	0.0119	1	0.1033	0.7986	0.4202
	0.0001	0.0001	0.4901		0.0001	0.0001	0.0001
researchers	0.0500	0.0792	0.0013	0.1033	1	0.0689	0.0904
	0.0038	0.0001	0.9364	0.0001		0.0001	0.0001
export	0.1601	0.8435	0.0624	0.7986	0.0689	1	0.3339
	0.0001	0.0001	0.0003	0.0001	0.0001		0.0001
patents	0.2778	0.3292	-0.0022	0.4202	0.0904	0.3339	1
	0.0001	0.0001	0.8988	0.0001	0.0001	0.0001	

p-value below coefficient.

• ***Innovation output:***

Innovation output (knowledge capital) defined as the ratio of sales of innovative products implies the direct association of the innovative activities of the firms with the performance of firms. Such association can be a key motivation of firms' innovative activities and also is related to their capacities of innovative activities.

KIS data contain three categories of innovation sales which is the ratio of innovative sales to total sales as follows: (i) products technologically new for the firm but not necessarily technologically new for the market, products partly or totally developed by the firm, and products introduced on the market, (ii) products technologically improved, and (iii) products technologically new both for the firm and for the market (radical innovations). Innovative output was calculated as the level of innovative sales after summing up the three components. The first two components are incremental innovations.

The innovation output is specified using the following exogenous variables: innovation input, level of productivity (or rate of productivity growth), human capital, co-operation on innovation, sources of knowledge for innovation, factors hampering innovation, industry classification and firm size.

• ***Productivity*:**

Labor productivity is defined as productivity per unit of labor, obtained as total sales divided by the number of employees in 2002.

• ***Other numerical variables*:**

Several other continuous or semi-continuous variables being included are patents, researchers, profit, and equity. Patents are intermediate output of technologically innovative activities and are associated with the level of developed technology innovation. Equity measures the capital intensity of the firm by dividing equity by the number of employees.

• ***KIS variables*:**

Several variables with high rate of responses from the 4 sub-samples such as firm innovation strategy, sources of knowledge for innovation, cooperation on innovation and factors hampering innovations are used in estimation as control variables. All these variables are ordered variables ranging from 0 (the lowest) to 6 (the highest).

In addition to the variables listed in the panels, three other variables such as product innovations (innovproduct), improved products (improveproduct) and *process* are used. Each of them corresponds to the number of innovations that firms carried out during the last 2 years for the development of the new innovative product, improved product and process innovations. These are also characterized as ordered variables.

• ***Dummy variables*:**

Firm size dummies, industry dummies, merger, closing, transfer and venture are among dummy variables used in the specification of the CDM model of Korean firms. Five size dummies and twenty one industry dummies are used in order to capture the firm and industry heterogeneities.

• ***Venture, Transfer, Merger and Closing:***

Venture is to indicate that firms are designated as a venture company or not. The variable transfer is to indicate whether firms have experienced the technology transfer from other companies or institutions. Merger and closing indicate whether mergers and closings have increased or decreased the firms' growth in sales by more than 10% per annum.

6. Empirical Model and Estimation Method

The econometric analyses of this study follow the basic ideas of CDM article. (1998) Our model consists of four equations, two of which are estimated by adopting Heckman's 2 step approach (Heckman, 1979), namely probit estimation in the first step followed by OLS estimation in the second step, and the last two equations of which are estimated in a simultaneous equation system framework. In estimating the system of equations, we have adopted three estimation methods such as instrumental variable (IV), two-stage least squares (2SLS) and iterated three-stage least squares (3SLS) methods to see whether the estimated elasticity of productivity with respect to innovation output is sensitive to the choice of estimation method.

The estimation process requires a number of steps. In the first step the innovation input equation is specified. In order to model identify the determinants of engagement in innovation activities and their impact on the probability of innovation investment, we estimate the innovation input equation firstly using a probit model. We use the obtained parameter estimates to calculate Mills Ratio indicating differences in probability of being innovative, which will be used in the second estimation step to correct for selection bias. In the second step, the innovation input equation is estimated using ordinary least square method based on only innovative sample. Here we estimate the effects of determinants of investment on the level of innovation investment.

The two models, that is, probit estimation and OLS estimation in the two step estimation, differ in the way that the dependent variable is defined. The dependent variable in the probit estimation is 1 if a firm is engaged in innovation investment and 0 otherwise – while in the least square estimation, it represents the level of innovation investment. Heckman's two step estimation can alternatively be estimated using a generalized tobit model accounting for selection bias. The later is estimated by maximum likelihood method.

6.1. Innovation Investment Equations

The innovation input equations are specified incorporating the following explanatory variables. Firm innovation is defined as R&D expenditures per employee in logarithmic term *(log(innovinput))* in 2001. The variables for R&D personnel in logarithmic term *(log(researchers)* and for profit in logarithmic term *(log(profit))* are included. We include Mill's ratio (*mratio*), reflecting differences in the probability of conducting innovation investments. In addition, it also includes two dummy variables *(Merger and Closing)* which indicate whether mergers and closings have increased or decreased the firm's growth by more than 10 percent. Innovation data from the Korea Innovation Survey (KIS) reflect: (i) innovation strategy, (ii) co-operation on innovation, (iii) sources of information for innovation, and (iv) factors hampering innovation. In order to capture the heterogeneity of firms, four size dummies are included in the specification.

The innovation input equation is written as:

$$\log(innovinput)_i = \beta_0^1 + \beta_1 \log(profit)_i + \beta_2 \log(researchers)_i + \beta_3 mratio_i + \beta_4 merger_i + \beta_5 closing_i + \beta_6 KIS_i + \beta_7 size_i + U_i^1 \quad (5)$$

6.2. System of Innovation Output and Productivity Equations

In the simultaneous equation system, Griliches (1994) describes the transformation process that goes from innovation input to economically valuable innovation output as the "knowledge production function". In this paper following Griliches we assume that the firm's ability to transform different kind of innovation input to innovation output is a function of the firm's accumulated knowledge capital as in (6):

$$\log(innovoutput)_i = \beta_0^2 + \beta_1 \hat{pinput}_i (+ \beta_2 \log(\hat{productivity})_i) + \beta_3 merger_i + \beta_4 closing_i$$
$$+ \beta_5 \log(patents) + \beta_6 transfer + \beta_7 venture + \beta_8 KIS_i + \beta_9 size_i + \beta_{10} ind_i + U_i^2 \quad (6)$$

The dependent variable, Knowledge Capital (innovation output), in equation (6) is measured as log of sales of innovative products *(log(innovoutput))*. Innovation output is determined by the predicted value of innovation investments, the number of patents, a variety of characteristics of innovative activities of Korean firms by KIS variables, technology transfer, merger and closing. In addition, since the simultaneous framework implies that we allow for feedback from productivity on innovation output, which suggests that innovation output and productivity performance are expected to be positively related, the predicted value of productivity was included for the IV estimation. Lastly, by introducing 21 industry dummy variables, the mean differences between industries, called industry effects, are captured, and by introducing 4 size dummies in the equation, firm effects are expected to be captured. Innovation output is estimated together with the productivity equation (7) in a simultaneous framework:

$$\log(productivity)_i = \beta_0^3 + \beta_1 \hat{poutput}_1 + \beta_2 \log(equity)_i + \beta_3 \log(researchers)_i$$
$$+ \beta_4 merger_i + \beta_5 closing_i + \beta_6 innovproduct_i + \beta_7 improvproduct_i + \beta_8 process_i + U_i^3 \quad (7)$$

Estimation in the last step is restricted to the sub-sample of firms with positive innovation input and innovation output. The dependent variable which is productivity in logarithmic term is measured by the total sales per employee. The productivity equation (7) is explained in part by predicted innovation output ($\hat{poutput}$) and capital per employee in logarithmic term ($\log(equity)$) and the number of product innovation and process innovation that firms carried out during the last 3 yeas. $(innovproduct, improvproduct, process)$. The remaining variables are accounted for the previous equations.

6.3. Estimation Methods

• *IV and 2SLS Estimation Methods:*

Using the ordinary least squares (OLS) estimation method to estimate these equations produces biased estimates. In order to solve this simultaneous bias, we estimated parameter by using predicted values of innovation output and productivity on the right-hand side. For the IV estimations, the predicted values were directly calculated by the authors and only one iteration was performed. On the other hand, the entire process of two-stage least squares was performed by the statistical package SAS.

• ***3SLS Estimation Method:***

3SLS is combination of the 2SLS and seemingly unrelated regression (SUR) methods to take into account both dependent regressors and cross-equation correlation of the errors. Residuals from the 2SLS method are used to estimate the variance-covariance matrix required for 3SLS. The model was also estimated by iterative 2SLS and 3SLS. To avoid excessive complexity the results are not reported here.

7. Empirical Results

7.1. The Innovation Input Results

Table 5 shows the results from the Heckman's two step estimation of innovation input model. We look first at the table for the decision to engage in R&D (probit column). Human capital measured by log of researchers is negatively associated with the probability of engaging in R&D activities. Also, E1A12 (Improve Product Quality), D1A1 (Sources Within the Enterprises), F1A11 (Excessive Perceived Economic Risks), F1A4 (Lack of Information Technology), the less likely the probability of engaging in R&D activities. By Contrast, the probability of engaging in R&D increases with G1A3 (Supplier).

Somewhat surprisingly, our results (probit column) do not show that the probability of engage in R&D activities increases significantly with the firm size (dummies assigned in the order of number of employees). This result is not consistent with 'Stylized Fact 1' stated by Cohen and Klepper (1996), which states that the likelihood of a firm reporting positive R&D effort rises with firm size and approaches one for firms in the largest size ranges' (p.928) However, the size 5 dummy which showed the highly significantly positive coefficient (at <0.001 level of significance) indicates large sized firms (number of employees>500) are more likely to engage in R&D activities compared to small sized firms (number of employees<10).

Now we turn to the Innovation Input (or R&D investment) equation (OLS column). In the second step of the model, the estimation of innovation input is based on only innovative firms having a different probability of innovative activities compared to non-innovative firms, a selection variable which is labeled as *mratio* was included to correct for the sample selection bias. The negative and significant parameter estimates showed the presence of selectivity in investment decision. The estimates show whether and to what extent the firms invest in innovative input, once they are engaged in innovative activities.

The coefficient of researchers is considered to be statistically significant. That is, once firms are engaged in innovative activities, the more researchers that firm hires, the more innovation investment firm is engaged in. However, profit in innovation input model is found to be insignificant. The status of a venture company is found to be significant to the innovation investment. It is expected that a venture company can be designated with high ratio of innovation investment.

Sale growth increased more than 10 percent due to merger with another firm or part of it is found to be positively significant. Looking at the estimates for the strategy variables, we find that investment in innovation is an increasing function of improving product quality and open up of new markets. In other words, the firms that desire to improve product quality and open up new markets appear to invest more in innovation. Regarding the sources of knowledge for innovation, sources within the enterprise (D1A1) is found to be negatively correlated with innovation investment. In contrast, when the sources of knowledge for innovation was reported to be from within universities (D1A16) or Computer-based information networks (D1A27), a positive association with innovation investment was found. However, all of those variables are found to be statistically insignificant.

Regarding Cooperation on innovation variables (G), Cooperation arrangements on innovation with domestic customers (G1A2) are found to be positively significant. On the other hand, cooperation with domestic suppliers (G1A3) is found to have a negative impact on the size of innovation investments. Cooperation with the domestic universities (G1A8) have positive coefficient but not significant. Regarding Factors hampering innovation the variables of Lack of qualified personnel (F1A3) and excessive perceived economic risks (F1A3) are found to positively affecting the innovation investment which are opposite from the expected signs. In sum, the characteristic of KIS data were hardly found significant or of appropriate sign for the analysis of innovation investment.

Our result shows that the Innovation Input (or R&D investment) of the firms engaging in R&D does not depend on the size in a systematic order, but the larger firms tend to invest more in R&D (size 4 i.e. number 250<employees<500 and size 5 i.e. number of employees>500) than the firms of the smallest size (number of employees<10). This result is somewhat consistent with Cohen and Klepper (1996), (See Cohen and Klepper (1996) for details of Stylized Facts 2 and 3).

Although we did not report the result in Table 5, we initially controlled for the different industries, which were 22 industries altogether. However, we did not find the heterogeneity in the probability of engaging in R&D activities and in R&D intensity by industries nor the heterogeneity in productivity, and also that adding the industry dummy altogether does not improve significantly the explanatory power of the models estimated. Hence, we exclude the industry dummy variables from model specifications outlined above.

7.2. The Innovation Output Results

The system of innovation output and labor productivity equation was estimated and the results are presented in Table 6 and 7. The two equations are estimated as system of interdependent equations by IV, 2SLS and 3SLS methods to check robustness of our results. The fit of the models in terms of coefficient of determination is highest in the 2SLS (R-square is 0.65).

The results show that the coefficient of innovation output in the labor productivity equation and the coefficients of labor productivity in the innovation out equations are highly statistically significant. This suggests a two-way causal relationship between innovation output and labor productivity and in support of 3SLS as appropriate estimation method. Thus we interpret our results based on 3SLS estimation and mention briefly about the results from other estimation methods whenever necessary.

Innovation output increases with firm size. Firms tend to have higher innovation output as they use a higher proportion of sources of knowledge for innovation from their enterprises (D1A1: Sources within the enterprises). Highly significant coefficient of *pproductivity*, predicted labor productivity, on innovation output implies that the higher the labor productivity of a firm, the higher innovation output measured by innovation sales in percentage of total sales.

Significantly positive association between Innovation output and the innovation input effort were found. It was statistically significant at 1% level. Patents also showed a positive and significant (at 1% significance level) association with Innovation Output in 2SLS. The firm effects captured by 4 size dummies are found to be positive at 1% significance level. The result was also very similar when estimated by 2SLS.

There were some differences in results depending on estimation method employed; especially regarding the characteristics of innovative activities of Korean firms we could not find consistently significant results. For example, *the firm strategy to open up new market* (E1A6) negatively contributes the innovation output in IV and 3SLS. On the other hand, *the strategy to reduced manufacturing costs* (E1A10) is found to be positively correlated with the innovation output in IV. Regarding the source of knowledge for innovation, *Sources within the enterprises* (D1A1) positively contributed to the innovative output throughout all approaches. The hampering factor such as *the excessive perceived economic risks* (F1A11) are found to be negatively significant in IV and 2SLS as expected.

We found the feedback effect of productivity on innovation output, which is statistically significant at 1% level. It is also significant in both IV and 3SLS indicating 2-way causal relationships between knowledge capital and productivity of Korean firms.

7.3. The Labor Productivity Results

In the productivity equation the 3SLS has a better fit than IV and 2SLS methods. As mentioned previously the results suggests a two-way causal relationship between labor productivity and innovation output. Thus, the analysis will primarily be based on the 3SLS method. However whenever necessary, we mention the difference or robustness across all models for the comparison purposes.

The results from the productivity presented in Table 7 show that productivity increases largely with knowledge capital. That is, the elasticity of innovation, which is predicted innovation output, obtained by instrumenting other variables on innovation output to labor productivity was positive. Also, its effects are statistically significant at 1% in all of the models. By contrast, the elasticity of human capital measured by researchers was negative and significant and also robust across all models. This result may raise a question on the quality of the researchers. Researchers and innovation output are correlated causing multicollinearity and problem of separation of their effects. The number of innovations which firm carried out in *products* and *process* is found to be negatively correlated with labor productivity in 2SLS. The elasticity of equity to labor productivity was positive and significant. Finally, our empirical result shows that the labor productivity is negatively associated with firm size and this result was robust across the 2SLS and 3SLS models.

Table 5. Heckman's Two Step (Probit/OLS) Estimation method

Probit of innovativeness			OLS of innovation input		
	Estimate	**P-value**	**Variable**	**Estimate**	**p value**
Intercept	1.3702***	<.0001	Intercept	3.4455***	<.0001
lresearchers	-0.6945***	<.0001	Lprofit	0.0124	0.2967
Merger	-0.1476	0.34	lresearchers	1.0078***	<.0001
Closing	-0.0639	0.7284	Mratio	-0.7583**	0.0427
E1A12	-0.0914***	0.0038	Merger	0.4102***	0.0074
E1A6	0.0109	0.6107	Closing	-0.0904	0.6364
E1A10	0.04	0.1458	venture	0.2967***	<.0001
D1A1	-0.0633***	0.0037	E1A12	0.0722*	0.0656
D1A16	-0.0121	0.6295	E1A6	0.0900***	<.0001
D1A27	-0.0109	0.6246	E1A10	-0.0689**	0.0227
G1A2	-0.0328	0.2842	D1A1	-0.0284	0.3218
G1A3	0.0856**	0.0124	D1A16	0.0025	0.9128
G1A8	-0.0196	0.5381	D1A27	0.0287	0.2074
F1A3	-0.0048	0.8991	G1A2	0.0505*	0.0879
F1A11	-0.1229***	<.0001	G1A3	-0.0835**	0.0431
F1A4	-0.1121***	0.0038	G1A8	0.0064	0.8255
size2	0.0531	0.7449	F1A3	0.0684*	0.0721
size3	0.1823	0.2735	F1A11	0.1017**	0.0235
size4	0.3267	0.1128	F1A4	-0.0126	0.7942
size5	0.8775***	0.0001	size2	0.1754	0.4976
			size3	0.4103	0.1241
			size4	0.6986**	0.0197
			size5	0.9628**	0.0129
Log Likelihood	-1270.14		R-Square	0.4253	
			Adj R-sq	0.4161	

Note) Dependent variable for tobit estimation is log innovation input per employee

Note) E: Firm innovation strategy, D: Sources of knowledge for innovation, G: Cooperation on innovation, and F: Factors hampering innovation

E1A12: improve product quality, E1A6: open up new market, E1A10: reduce other manufacturing costs, D1A1: sources within the enterprises, D1A16: universities, D1A27: computer-based information networks, G1A2: customers, G1A3: supplier, G1A8: universities, F1A3: lack of qualified personnel, F1A11: excessive perceived economic risks, F1A4: lack of information technology

Table 6. System of equation
[Dependent variable is innovation output measured by log of innovative sales]

	IV Estimates		2SLS		3SLS	
Variable	Estimate	p value	Estimate	P value	Estimate	p value
Intercept	1.2798	0.1512	0.2130	0.6953	-1.0837**	0.0151
pproductivity	0.7117***	0.0001	0.2952***	0.0001	0.2871***	0.0001
Pinput	0.3983***	0.0001	1.0177***	0.0001	1.2367***	0.0001
Merger	0.0483	0.7815	0.0710	0.5913	0.0642	0.6232
Closing	0.1127	0.6121	0.0956	0.5695	0.1768	0.2857
Transfer	0.0826	0.3732	0.1249*	0.0758	0.0323	0.3949
Lpatents	0.0663*	0.0656	0.1006***	0.0002	0.0200	0.182
size2	-0.0839	0.7417	1.2086***	0.0001	1.1271***	0.0001
size3	-0.2714	0.4023	2.1560***	0.0001	2.0508***	0.0001
size4	-0.3957	0.3195	2.7978***	0.0001	2.7017***	0.0001
size5	-0.6592	0.1736	3.5876***	0.0001	3.4516***	0.0001
E1A12	-0.0369	0.4503	-0.0290	0.4349	-0.0063	0.752
E1A6	-0.0612**	0.0248	-0.0176	0.4036	-0.0291**	0.0172
E1A10	0.0637*	0.0573	0.0409	0.1098	0.0243	0.0811
D1A1	0.0660**	0.0127	0.0843***	0.0001	0.0317***	0.0047
D1A12	0.0065	0.7922	-0.0197	0.301	0.0005	0.9569
D1A16	-0.0169	0.5367	0.0052	0.8032	-0.0027	0.8043
D1A27	-0.0352	0.197	-0.0283	0.1733	-0.0128	0.2506
G1A2	-0.0191	0.5537	-0.0016	0.9461	-0.0057	0.6639
G1A3	0.0444	0.2454	0.0096	0.7402	0.0111	0.4755
G1A8	0.0031	0.9251	0.0283	0.2698	0.0031	0.8166
F1A3	-0.0541	0.229	-0.0415	0.223	-0.0216	0.2364
F1A11	-0.0722**	0.0372	-0.0626**	0.0173	-0.0203	0.1582
F1A4	0.0282	0.5373	-0.0039	0.9093	0.0161	0.3915
R-square	0.5262		0.6693		System Weighted R-Square	0.4529
Adjusted R-square	0.5053		0.6547			

Note1) *, **, *** denote significance at the less than 10%, 5%, and 1% levels

Note2) 21 industry dummies included in OLS, 2SLS and 3SLS.

Note 4) E: Firm innovation strategy, D: Sources of knowledge for innovation, G: Cooperation on innovation, and F: Factors hampering innovation

E1A12: improve product quality, E1A6: open up new market, E1A10: reduce other manufacturing costs, D1A1: sources within the enterprises, D1A16: universities, D1A27: computer-based information networks, G1A2: customers, G1A3: supplier, G1A8: universities, F1A3: lack of qualified personnel, F1A11: excessive perceived economic risks, F1A4: lack of information technology

Table 7. System of equations
[Dependent variable: log of productivity per employee]

	IV Estimates		2SLS		3SLS	
Variable	**Estimate**	**p value**	**Estimate**	**p value**	**Estimate**	**p value**
Intercept	1.7734***	0.0005	0.6110*	0.0565	0.4018	0.1918
linnovoutput	0.3546***	<.0001	0.6571***	<.0001	0.7274***	<.0001
lresearchers	-0.0891**	0.0144	-0.1431***	<.0001	-0.1335***	<.0001
Lequity	0.1893***	<.0001	0.1056***	<.0001	0.0301***	0.0098
Merger	-0.0661	0.579	-0.1060	0.2745	-0.1187	0.2205
Closing	-0.1647	0.2671	-0.1339	0.2746	-0.1114	0.3628
innovprod	-0.0017	0.9404	-0.0442**	0.0281	-0.0114	0.3074
improvprod	0.0253	0.2287	-0.0417**	0.0238	-0.0062	0.5497
Process	-0.0131	0.5717	-0.0650***	0.0011	-0.0151	0.1811
size2			-0.5746***	0.0048	-0.8038***	<.0001
size3			-1.1363***	<.0001	-1.5089***	<.0001
size4			-1.5238***	<.0001	-2.0231***	<.0001
size5			-2.0977***	<.0001	-2.6723***	<.0001
R-square	0.1548		0.2766			
Adjusted R-square	0.1447		0.2680		R-Square	0.4529

Note) *, **, *** denote significance at the less than 10%, 5%, and 1% levels

Table 8. **Description of KIS data:**
Ordered variables which characterize innovation activities of Korean firms

Panel A. Firm innovation strategy by its degree of importance.

Strategy	**All observations (n=3331)**		**Innovative sample (n=1182)**	
(unit:%)	**S.I.**	**Q.I.**	**S.I.**	**Q.I**
Product innovations				
Open up new markets	15.22	10.96	28.17	19.77
Improve product quality	29.12	13.93	50.93	25.21
Replace products being phased out	21.07	14.44	38.32	26.40
Extend the product range	16.84	16.72	30.29	31.13
Fulfill regulations, fulfill standards	9.055	10.36	41.665	23.815
Process innovations				
Reduce labor cost	16.99	12.55	26.73	22.25
Improve production flexibility	10.21	11.56	17.68	20.90
Reduce other manufacturing costs	20.14	14.44	33.16	27.24

Table 8. continued

Panel B. Sources of knowledge for innovation by degree of importance

Strategy	All observations (n=3331)		Innovative sample (n=1182)	
(unit:%)	S.I.	Q.I.	S.I.	Q.I
Customers	17.59	12.64	31.64	23.77
Sources within the enterprise	17.13	11.67	31.74	21.43
Competitors	10.39	12.67	18.87	23.69
Sources within the group	3.33	5.07	5.75	8.63
Fairs, exhibitions	9.04	12.55	16.67	24.28
Suppliers	6.12	10.075	9.945	17.385
Professional conferences, meeting, journals	4.08	9.46	7.78	19.20
Universities	3.87	6.24	7.36	13.20
Patent disclosures	6.90	8.35	14.38	17.94
Consultants	1.92	4.08	7.78	19.20
Computer-based information networks	7.87	13.06	14.81	25.21

Panel C. Co-operation on innovation

	All observations	Innovative sample
Customer	12.52	24.03
Supplier	10.18	18.485
Universities	8.92	18.95
Firms within the group	7.39	13.03
Government or non-profit research institutes	7.36	15.74
Consultancies	4.8	9.48
Competitors	7.02	12.94

Panel D. Factors hampering innovations

	All observations	Innovative sample
Lack of qualified personnel	41.1	73.18
Excessive perceived economic risks	31.64	57.53
Organizational rigidities	18.55	31.9
Innovation cost too high	30.02	54.82
Lack of appropriate sources of finances	29.33	52.79
Lack of customer responsiveness	29.03	51.86
Lack of information on technology	39.03	69.46
Lack of information on markets	33.86	61.17
Fulfilling regulations, standards	22.205	39.81

Note1) The responses to panel A, B, C and D are indicated by six categories in the questionnaire such as strongly, quite, moderate, weak and not important and not relevant

Note2) S.I. indicates strongly important and Q.I indicates quite important

Note3) Values in panel C and D indicate the percentage of firms of which response correspond to strongly, quite and moderately important to the questions.

8. CONCLUSION

This study was aimed at investigating the relationship between innovativeness and productivity of Korean manufacturing firms. In doing so, we have examined the possible two-way causal relationships between knowledge capital and productivity among the Korean innovative manufacturing firms. We have used a *knowledge production function* approach formulation of the link between incentive to invest in innovation activities, innovation input, innovation output and labor productivity growth by Pakes and Griliches (1984). The models refined by by Crepon, Duget, and Mairesse (1998) are estimated by adopting a multi-step econometric approach suggested in Lööf and Heshmati.

Our model consists of 4 equations; First two equations, Innovation Investment Decision and Innovation Input equations were estimated by Heckman two-step approach to control for sample selection bias of estimating only those who engage in innovative activities. In the two-step method, we used probit model to estimate the Innovation Investment Decision equation and calculated Mill's ratio which was used as an explanatory variable in the second equation, Innovation Input Equation, to control for the selectivity bias. The second step of the estimation was done by least squares estimation (OLS) applied only to sample of innovative firms.

The last two equations, Innovation Output equation and Productivity equation were estimated as a system. Estimating the system of equation required us to control for simultaneity bias because we allowed productivity to be an explanatory variable of Innovation Output equation and Innovation Output variable to be an explanatory variable of Productivity equation in order to investigate the possible two-way causal relationship between the two variables. To see the robustness of the estimates, the model was estimated with several estimation methods, namely IV, 2SLS and 3SLS.

For the empirical analysis we have used Korean Innovation Survey data released in 2002, which contains detailed information on various key characteristics of Korean manufacturing firms related to innovation characteristics. After controlling for firm specific variables that are related to firms' innovation activities such as innovation strategy, innovation sources, sources of cooperation on innovation and the innovation hampering factors we have found that the firm size does not influence firms' decision to engage in innovation activities nor the innovation input intensity once they decide to engage in innovation. On the other hand, our study found that the firm size was negatively related to the labor productivity in Korean manufacturing sector. Furthermore, some firm-specific effects that positively contribute to innovation output found to be negatively related to productivity, suggesting that the factors that enhance the innovativeness of firms may not enhance their productivity growth.

Finally, our findings suggest that Korean innovative firms are subject to two-way causal relationship between knowledge capital and labor productivity. This is interpreted as investment in innovation affects positively labor productivity of firms, which in turn result in more investment in innovation through feedback effect from labor productivity growth. Industry heterogeneity was not found to affect innovation output nor labor productivity of innovative firms in the Korean manufacturing industry.

APPENDIX A. OVERVIEW OF KOREAN INNOVATION SURVEY DATA

Korea Innovation Survey (KIS) has been undertaken twice since 1995 by the Science and Technology Policy Institute (STEPI) based on the OECD Oslo Community Innovation Survey (CIS) manual. This dataset which comprises data about innovation activity at the firm level in Korea covers more than 3,775 firms and 21 industrial sectors based on the Hankyung Database (DB) industrial classification. This CIS 2002 used in current study contains detailed questions on various innovation activities at the firm level, which have taken place between January 2000 and December 2001 at firm level. The survey is based on the manufacturing firms who employ at least 10 people or greater and who were founded before year 2000. Based on this population (however, tobacco and recycling industry, that is Korean Standard Industrial Classification (KSIC) 15 and 46, are excluded), the sample is stratified.

For the investigation, the survey questionnaire includes ten categories of survey parameters, such as (1) general information about the business operation, (2) innovative performance of the firm, (3) information sources of innovation, (4) purposes of innovation, (5) innovation cost, (6) impacts of innovation, (7) technology acquisition and transfer, (8) protection of intellectual property rights, (9) assessment of institutional/governmental support for innovative activities, and finally (10) obstacles to innovation. The measures of survey parameters include quantitative and qualitative ones.

The definition of technological innovation can vary. In the Community Innovation Survey, technological innovation refers to 'invention of technologically new products or process, or technological improvement either in product or in process.' According to the definition, 'innovative firms' refer to the firms who have been successful either in product innovation or process innovation at least once during the underlying period. This term differs from 'innovative active firm' which refer to the firms who have been conducting the technological innovation during the period regardless of whether the innovation has been successful or not. That is, 'innovative active firm' include not only the innovative firms but also the firms who have tried innovation but happened to fail or whose innovation has been incomplete. Those who have not conducted any innovative activity is referred to as 'non-innovative firms' (STEPI 2002).

In this survey, new product innovation implies a technologically new product whose technological characteristics or intended uses differ significantly from those of previously produced products. The product improvement implies a technologically improved (existing) product whose performance has been significantly enhanced or upgraded. The process innovation implies the adoption of technologically new or significantly improved production methods, including methods of product delivery. These methods may involve changes in equipment, or production organization, or a combination of these changes, and may be derived from the use of new knowledge.

Innovation activity in Community Innovation Survey is restricted to the activities that are introduced in the market and thus results in an increase in sales or reduction in costs. If the proportion of innovation active firms is high among the entire firms of a country, the capability of technological innovation in that country, especially in the particular sector, can be assessed to be high

The characteristics of technological innovation of Korean firms are overviewed in Table 8. Panel A shows the firm's strategy on innovations distinguished by the degree of importance. The most important objectives of product innovations are improvement in product quality and replacement of the products being phased out. Fulfilling regulations and fulfilling standards dominate among the most important objectives when the strong importance answers of the innovative sample are considered. Looking at the most important sources of knowledge for innovation (Panel B), we find that customers and sources within the firm dominate completely when strong importance is considered. Domestic customers and suppliers and universities are the most common co-operative partners in innovation. Panel D presents some summary statistics considering factors negatively affecting innovation. Overall, the higher rate of this panel compared to those of other panels are pronounced. The high rate of this panel implies that Korean firms have a strong intention to make innovation but there exist many factors which hinder such innovative activities. A lack of qualified personnel and lack of information for technology were identified as the two most important factors delaying innovation. The degree of importance is much higher for the sample of innovative firms. However, the ranking of factors does not change when the two samples are compared.

References

Bartelsman, E., & Doms, M. (2000). Understanding productivity, lessons from longitudinal micro datasets, *Federal Reserve Board Working Paper*, Washington, DC.

Branscomb, L.M. & Choi, Y-H. (1996). Eds, *Korea at the Turning Point: Innovation-Based Strategies for Development*, Westport, Connecticut, London: Praeger.

Cohen, W. & Klepper, S. (1996). A Reprise of Size and R&D, *The Economic Journal,* **106**, 925-951.

Cohen, W. M. & Levin, R.C (1989). Empirical studies of innovation and market structure, In R. Schmalensee & R.D. Willig, *Handbook of Industrial Organization* 2 (pp. 1060-1107) B.V., Amsterdam, Elsevier Science Publishers.

Crepon, B., Duget, E. & Mairesse, J. (1998). Research, innovation, and productivity: An econometric analysis at the firm level, *Economics of Innovation and New Technology*, **7** (2), 115-158.

Foster, L., Haltiwanger, J. & Krizan, C.J. (1998). Aggregate productivity growth: Lessons from microeconomic evidence, *NBER Working Paper* No. **6803**, Cambridge, MA.

Griliches, Z. (1994). Patent statistics as economic indicators: A survey, *American Economic Review* 84, 1–23.

Heckman, J. (1979). Sample Selection Bias as a Specification Error, *Econometrica* **47**_(1): 153-161.

Jeong, K-Y. & Masson, R.T. (2003). A new methodology linking concentration dynamics to current and steady-state profits: Examining Korean industry policy during take-off, *International Journal of Industrial Organization* **21**, 1489-1526.

Klette, J, & Griliches, Z. (1998), Empirical patterns of firm growth and R&D investment: A quality ladder model interpretation, *NBER Working Paper* No. **6753**. Cambridge, MA.

Lee, J.S, Kim, D.H. & Kim, S.H. (2004). Firm Effect versus Industry Effect on Firm R&D Intensity, *STEPI Innovation Survey Workshop Working Paper*, STEPI, Seoul, Korea

Lööf, H. & Heshmati, A. (2002). Knowledge capital and performance heterogeneity: A firm-level innovation study, *International Journal of Production Economics* **76**, 61-85.

Mairesse, J. & Sassenou, M. (1991). R&D and productivity: A survey of econometric studies at the firm level, in: *Science-Technology, Industry Review*, **8**, OECD, Paris, 9–43.

OECD, 2006, Gross Domestic Product per Capita for OECD Countries, Retrieved February 9, 2006, from OECD Web Site: http://www.oecd.org/dataoecd/48/5/34244925.xls

Pakes, A. & Griliches, Z. (1984). Patents and the R&D at Firm Level: A First Look, In Z. Griliches, *R&D, Patents and Productivity* (pp. 390-409). Chicago: University of Chicago Press.

Roller, L.H. & Sinclair-Desgagnle, B. (1996). Industrial organization and business strategy on the heterogeneity of firms, *European Economic Review* **40**, 531–539.

Sakakibara, M. & Cho, D-S. (2002). Cooperative R&D in Japan and Korea: a comparison of industrial policy, *Research Policy* **31**, 673-692.

Seo, J-H, (2004). *Comprehensive Study on Industrial Competitiveness of Korea* (2). Seoul, Korea: Korea Development Institute.

Seo, J.W. (2004). Analysis on Difference in Effects of Multiple Sources of Technological Innovation, *STEPI Innovation Survey Workshop Working Paper*, STEPI, Seoul, Korea

Seo, K.W. & Lee, C.Y. (2004). The hampering factors of technological innovation in Korean firms, *Technological Innovation Study,* **12**(1)

Um, M.J et al. (2002), *Korean Innovation Survey*, Seoul, Korea: STEPI

Sung, T.K, (2004). Networks, Technological Opportunity, and Innovative Activity: Evidence from the Korean Manufacturing Firms, *STEPI Innovation Survey Workshop Working Paper*, STEPI, Seoul, Korea

In: Economics of Technological Change
Editor: Thomas L. Beane, pp. 69-88

ISBN: 978-1-60021-237-6
© 2007 Nova Science Publishers, Inc.

Chapter 4

THE IMPACT OF UNIVERSITY RESEARCH AND DEVELOPMENT ON AUSTRALIAN REGIONAL DEVELOPMENT

Claudia Burgio-Ficca and Hristos Doucouliagos[1]
School of Accounting, Economics and Finance,
Faculty of Business and Law, Deakin University

ABSTRACT

Recent years have seen growing interest in research and development (R&D) in Australia. Unfortunately, little is know about the contributions of R&D to Australian economic performance. Even less is known about the contributions of university R&D. This paper examines the impact of both university and non-university R&D on output levels in Australia, using data for Australian states, for the period 1979 to 2003. Both Cobb-Douglas and Translog versions of the production function are estimated. The results indicate that while both types of R&D make important contributions to the generation of output, not all university R&D has a positive effect. Pure basic and applied university R&D both have had a positive effect on regional output. However, this is in sharp contrast to the impact of strategic basic and experimental R&D, which are shown to have a negative effect on output.

I. INTRODUCTION

The idea that research and development (hereafter referred to as R&D) is an important and vital input into the production process has been emphasized by economists for at least the last 40 years. Internationally, the topic of R&D and its impact on productivity and growth has been extensively explored and documented. Studies by Mansfield (1972), Nadiri (1980), Griliches (1992) and Coe and Helpman (1995), to name but a few, have all concluded that R&D expenditure contributes substantially to the growth of output. However, reviews by

[1] E-mail address: douc@deakin.edu.au. Mailing address: School of Accounting, Economics and Finance, Faculty of Business and Law, Deakin University, 221 Burwood Highway, 3125, Victoria, Australia. (Corresponding author.)

Dowrick (1994) and the Industry Commission (1995) indicate that Australian studies in this area are limited and have largely been confined to the agricultural and manufacturing industries. Moreover, most of the extant Australian research has been in the form of case studies, as opposed to econometric studies.

In recent years, the Commonwealth Government has commissioned a number of reports into the current and future status of R&D in Australia. These reports have included *The Virtuous Cycle: Working together for health and medical research* (Health and Medical Research Strategic Review, 1999), *Knowledge and Innovation: A policy statement on research and research training* (Kemp, 1999), *The Chance to Change* (Batterham, 2000) and *Backing Australia's Ability* (Commonwealth of Australia, 2001). While these reports provide a review of current funding levels for various fields of research, and have proposed a number of new initiatives for the future of R&D, they have done little to analyze the current impact R&D has on Australian output. Such an assessment is long overdue and we hope to add to this, even in a small way, in this paper.

In Australia, the greatest proportion of R&D comes from the private sector, with the university sector being the second greatest contributor, followed by the government sector in third place. Even so, in 2002-03, most (about 86%) of university R&D was funded by the Commonwealth Government, with about 5% of the funding coming from the private sector (Year Book Australia 2005).

This paper's main aim is to examine the impact of university R&D on Australia's development, as proxied by output. This is achieved by estimating production functions using a panel data set of six Australian states for the period 1979 to 2003. The paper is set out as follows. Section II provides a brief review of previous research in this area. Section III outlines the estimation procedure undertaken, while Section IV discusses the data. Section V presents the results. Concluding comments are made in Section VI.

II. Previous Research

Early econometric analyses linking R&D to productivity were conducted by Terleckyj (1958), Minasian (1962), Griliches (1964) and Mansfield (1965). Terleckyj (1958) used the total factor productivity (TFP) approach, relating R&D intensity to the rate of growth in the residual output (controlling for the contributions of labor and physical capital). Terleckyj concluded that the intensity of R&D may contribute to productivity. Minasian (1962) also used the TFP approach and found that R&D expenditure was a highly significant independent variable in determining the rate of growth in productivity in a cross-section of firms in the chemical and allied products industries and in the drug and pharmaceutical industries. Griliches (1964) used an unrestricted Cobb-Douglas production function and concluded that public expenditure on agricultural research had a positive and significant affect on agricultural output.

Since these early works, there has been an abundance of studies exploring the impact of R&D on productivity. Noteworthy contributions were made by Griliches (1979) and Jaffe (1986). Applications of the TFP method include Goto and Suzuki (1989), Griliches (1994) and Sterlacchini (1989). Some of the more important reviews of this literature are provided by Mairesse and Sassenou (1991), Nadiri (1993), the Industry Commission (1995), Mairesse and Mohnen (1994) and Mohnen (1996).

Unfortunately, the impact of university research on productivity has received only scant attention, in comparison to the wider attention of R&D in general. The existing research focuses mainly on the existence of spillovers from university R&D and the impact of university R&D on innovation. These are important *indirect* channels through which university R&D can affect output. However, very little is known about the *direct* effect of total university R&D on output. Even less is known about the direct impact of different types of university R&D.

An early study on university R&D was conducted by Jaffe (1989). Jaffe's analysis aimed to examine the production of corporate patents and how this was influenced by industry and university R&D. The model included 29 U.S. states over the period 1972-1981 and incorporated a breakdown of university research by academic departments. Jaffe used a Cobb-Douglas specification estimated by both OLS and Generalized Least Squares (GLS). GLS produced better results for both the industry R&D and university R&D variables. Overall, Jaffe found that university expenditure on drugs and medical technology produced the statistically strongest effect, while chemical technology ranked second, followed by electronic, optical and nuclear technology. Importantly, Jaffe's study found that there was only a weak relationship between the location of the R&D spillovers and the location of the university and research laboratories, and that spillovers of R&D were limited to specific research areas. Anselin, Varga and Acs (1997, 2000) later disputed the findings on the location of R&D spillovers and the location of university and research laboratories.

Commenting on Jaffe (1989), Acs, Audretsch and Feldam (1992) argue that: "the knowledge created at university laboratories spills over to contribute to the generation of commercial innovations by private enterprises". Acs, Audretsch and Feldam (1994a) examined the degree to which university and corporate R&D spills over to innovative activity at the state level. Their analysis was based on the same data set as Jaffe (1989), however instead of investigating the contribution of university R&D to corporate patents, innovations were introduced as the dependent variable. They found that substituting patent data with a measure of innovative activity generally strengthened Jaffe's (1989) arguments and findings. In fact, unlike Jaffe (1989), Acs, Audretsch and Feldam (1992) conclude that spillovers were affected by the geographic location of universities and research laboratories.

Acs, Audretsch and Feldam (1994b) examined the degree to which university and corporate R&D spills over into firms. They concluded that R&D undertaken by university laboratories spilled over and impacted upon the innovative activity of small firms, whilst corporate R&D affected larger firms. Anselin, Varga & Acs (2000) extended their 1997 paper in which spillovers generated by universities was disaggregated at the sectoral level. Their findings indicate that local university spillovers were constrained to specific industries.

The link between research and innovation was examined also by Nelson (1986). Using a survey of research managers, Nelson concluded that university research was an important source of innovation in some industries. Link & Rees (1990) compared the university based research relationship between small and large firms in an attempt to explain why small firms are more innovative. Their analysis of 209 innovating firms concluded that larger firms were in fact more active in university based research, but were unable to exploit this association to the degree achievable by smaller firms. Anselin, Varga and Acs (1997) used the number of high technology innovations as their dependent variable, and found a positive and statistically significant relationship between university research and innovative activity.

In recent years various authors from the Melbourne Institute of Applied and Social Research have contributed papers to a collaborative research program. For example, Fenney and Rogers (2001) compared the performance of firms relative to their innovative activity. Their results show that innovation leads to increased firm performance. Using a cross-country regression approach, Rogers (2001) examined the economic growth of 61 countries between 1965 and 1985. With the trend growth in GDP per worker as the dependent variable, Rogers added each of the absorptive capacity proxies (20 in total) individually to the baseline regression. The absorptive capacity proxies were grouped under 5 main categories which comprised students studying abroad, telecommunications (i.e. main telephone lines per population), publications (i.e. periodicals published), patents (per thousand worker) and trade data (i.e. imports/export variables). The overall results suggest that absorptive capabilities (i.e. patents etc) affected the determination of economic growth.

Rogers (2002) analysed how innovative activity affected firm profitability in a sample of Australian firms. The study examined innovative activity from 1994 to 1998. The results revealed that patenting activity was never significantly associated with market value, implying that either patent applications were a poor proxy for innovative activity or more important that the stock market failed to recognise such value.

Unfortunately, none of these industry studies explores the impact of R&D on aggregate sectoral output. It is this lacuna that we wish to address.

III. Estimation Procedure

Our analytical approach is to estimate a set of production functions. The production function relates the quantity of output to the inputs used in the production process. In this framework, labor and capital are mandatory. In following with previous studies, research capital is introduced as a third input (see, for example, Coe and Helpman 1995, Griliches and Mairesse 1984 and Nadiri 1980). The Cobb-Douglas version of the basic production function is given by:

$$\ln(Y_{it}) = \alpha_0 + \alpha_1\ln(L_{it}) + \alpha_2\ln(K_{it}) + \alpha_3\ln(TRD_{it}) + \alpha_z\mathbf{Z}_{it} + u_i \tag{1}$$

where Y is output (in our case Gross State Product), L is labor, K is the stock of physical capital, TRD is the stock of total R&D capital, **Z** is a vector of other variables, u are the residuals and the subscript i is an index for each state, at time period t.

Separating higher education R&D from non-university R&D results in equation 2:

$$\ln(Y_{it}) = \gamma_0 + \gamma_1\ln(L_{it}) + \gamma_2\ln(K_{it}) + \gamma_3\ln(NURD_{it}) + \gamma_4\ln(URD_{it}) + \gamma_z\mathbf{Z}_{it} + v_i \tag{2}$$

where NURD is R&D capital from the non-university sector (mainly the private sector), and URD is university R&D capital. Equation 2 can be extended by separating out the different types of university R&D:

$$\ln(Y_{it}) = \delta_0 + \delta_1\ln(L_{it}) + \delta_2\ln(K_{it}) + \delta_3\ln(NURD_{it}) + \delta_4\ln(P_{it}) + \delta_5\ln(S_{it}) + \delta_6\ln(A_{it}) + \delta_7\ln(E_{it}) + \delta_z\mathbf{Z}_{it} + \varepsilon_i \tag{3}$$

where the stock of higher education R&D is disaggregated into four major categories: pure basic (P), strategic basic (S), applied (A) and experimental (E)[2]. It is reasonable to assume that given a very high degree of factor mobility and competition, that the production process is the same across each Australian state. Hence, it is assumed that all the variables have the same coefficient in each state, as well as over time[3]. The Translog versions of equations 1 to 3 include interactive terms.

Equations 1, 2 and 3 assume errors that are free of technical inefficiency – the production process is assumed to be fully efficient. In preliminary analysis, we estimated various Stochastic Production Frontiers. In no case was technical efficiency identified.[4] Equations 1 to 3 do not include a time trend, which is usually incorporated to represent disembodied technical change. In this paper, R&D is used instead of a time trend, as R&D is a broader measure that captures also technical change, and this is preferable to a simple time counter[5].

It is important to note that our focus in this paper is on only one aspect of the university research process - R&D. The impact of scholarship, communication of research findings and research training in general, especially higher degrees by research degrees, may all make important contributions to output, that are not included in the above specifications and are, hence, not considered in this paper. Moreover, knowledge created by universities can affect a society's welfare through its wider impact on education and culture. There are clearly many possible extensions to this paper.

Numerous factors affect the performance of the Australian economic landscape such as business cycles, government policies, regulatory reform, general world economic conditions, and technological change. Ideally one would like to account for all of these factors, so that the regression results provide the truest reflection of the impact on various variables on the Australian output. However, data and measurement issues prevent the vast majority of such factors being accounted for directly. We do however include two important control variables as part of the **Z** vector: the effect of industrial relations and deregulation on Australia's production process.

Unions and industrial relations issues in general have been a key feature of the industrial landscape in Australia. For example, the Australian Council of Trade Unions (ACTU) has had a significant impact on economic and social policy in Australia, both through its own actions and through its links with the Australian Labour Party (Brambles, 2001 and Briggs, 2001). There exists a voluminous theoretical and empirical literature on the economic effects of unions (see, for example, Turnbull 2003, Addison and Belfield 2004 and Hirsch 2004). In general, unions may be viewed as a having 'two-faces' (Freeman & Medoff, 1984). Unions are able to increase wage levels and impose employment restrictions. However unions are also able to increase productivity levels through strengthening communication between management and employees. The literature has failed to draw firm conclusions regarding the impact of unions. A recent meta-analysis of the literature concluded that for the U.S. there was a small positive effect on output levels, while for the U.K. the effect of unions was

2 A definition of each category is provided below in Section IV.

3 Similar assumptions were made by Acs, Audretsch & Feldman (1994a, 1994b), Anselin, Varga and Acs (1997), Griliches (1964) and Jaffe (1989).

4 This does not, however, mean that Australian states are not inefficient relative to other countries or regions in other countries.

5 Preliminary results revealed that the inclusion of a time trend caused both the time trend and the R&D variable to become statistically insignificant. As a result the time trend was omitted from the analysis. Similar problems were encountered by Nadiri (1980).

negative (Doucouliagos and Laroche 2003). A small group of studies has explored the impact of Australian unions (e.g. Alexander & Green: 1992, Drago & Wooden: 1992, Drago, Wooden & Sloan: 1992 and Phipps and Sheen: 1994). The verdict appears to be that Australian unions have had an adverse effect on production.

Another area of interest in the Australian context is the impact of deregulation of the financial system on the production process. This has been an extensive process, involving the floating of the Australian dollar in 1983, followed by the removal of a number of regulatory measures (e.g. exchange rate controls and deregulation of the banking sector), together with the privatisation of numerous government corporations. These events have significantly impacted upon the structure of the economy and as such their inclusion in the production function is warranted. Most of the extant research in this area has focussed on the impact of deregulation on specific industries (e.g. Doucouliagos & Hone 2000 and Lie, Brooks & Faff 2000), although there have also been some studies exploring the broader impact on the economy (Groenewold 2003). We expect to find that deregulation has had a positive effect on production.

IV. Data

The data used in this analysis was primarily sourced from the Australian Bureau of Statistics (ABS) for the six Australian states of New South Wales (NSW), Victorian (VIC), Queensland (QLD), Western Australia (WA), South Australia (SA) and Tasmania (TAS) between 1979[6] and 2003. Due to a lack of data, the regions of Northern Territory and ACT were eliminated from the analysis. Hence, this paper constructs a panel data analysis of six states and a twenty-five year span, resulting in 150 observations.[7]

The definitions of the variables used are as follows:

Output: While output levels are only one aspect of development, it is certainly an important one. We use the level of output at the state level as our broad measure of regional development. We used two measures of output. First, we used Gross State Product (GSP) deflated into constant 2002-03 dollars using the associated Gross State Product deflator. Second, we also used the Gross State Product Chain Volume Index series. The results are similar and we focus here on the results using the Chain Volume Index series.

Labor: measured as the total number of workers employed in each state.

Union Density: measured as the proportion of the workforce who are trade union members.

Deregulation: In the last two decades the Australian economy has experienced a process of deregulation in various industries through tariff reductions and industry protection. As there is no index available on deregulation in Australia, the focus of this paper is on the post-1983 period during which the majority of the deregulations were implemented. Hence, events such

6 1978-89 is the initial year the Australian Bureau of Statistics commenced collecting R&D expenditure data.

7 It is important to note that the Australian Bureau of Statistics provides data on R&D expenditure in Australia only biannually. In the absence of complete data, an average of the two years was taken for missing years data. A similar methodology was utilised by Olley and Pakes (1996).

as the floating of the Australian dollar and financial deregulations are accounted for through a dummy variable equivalent to zero for the years between 1979 and 1983 and one for the years 1984 to 2003. An alternative measure is to use the Fraser Institute measure of economic freedom (Gwartney and Lawson 2003). The results regarding R&D are not altered when this measure is used.

R&D Stock: In Australia, R&D is conducted by a number of institutions and governing bodies, including business enterprises, general government, higher education institutions and private non-profit organisations. The focus of this paper is on R&D conducted by higher education organizations (URD), compared to that arising from the other sectors (NURD). R&D has been defined as: 'creative work undertake on a systematic basis in order to increase the stock of knowledge – including knowledge of man, culture and society – and the use of this knowledge to devise new applications'[8]. However, what is of more importance for our purpose is the *stock* of knowledge generated by the R&D. In the absence of a stock series from the ABS, stock data[9] was constructed for the required variables: physical capital, total R&D, higher education R&D and the four various types of higher education R&D (pure, strategic, applied and experimental).

The stock of R&D for each of these variables was calculated using the perpetual inventory method (PIM) (see Coe and Helpman 1995). Under the PIM, R&D capital stock in any period is a function of the value of the R&D capital stock in the previous period, less any depreciation that has occurred plus the level of R&D expenditure in the previous period[10]. The PIM also requires the calculation of an initial stock of capital which is based on the growth of R&D and the assumed rate of depreciation[11]. Like all capital, the stock of accumulated knowledge from R&D expenditure depreciates (Mohnen, 1990). In this paper, it is assumed that R&D expenditure depreciates at the rate of 10 percent per annum. According to Griliches (1979), a rate of 10 percent is appropriate given that most private knowledge becomes obsolete after ten years. This rate was also adopted by Mohnen, Nadiri and Prucha (1986) in their analysis of manufacturing sectors in the US, Japan and Germany, and by

---

8 OECD (1980).

9 Stock data is a more accurate measure as it takes into consideration the depreciable/obsolete nature of R&D and provides a more accurate representation of the R&D available as an input into the production process.

10 $$S_t = (1-\delta)\, S_{t-1} + R_{t-1}$$

where: S_t = the stock of R&D capital at the beginning of the period

S_{t-1} = the stock of R&D capital at the beginning of period t-1

R_{t-1} = the expenditure on R&D during period t-1

δ = the depreciation or obsolescence in the rate of knowledge.

11 $$S_o = R_o/(g + \delta)$$

where: S_o = the stock of R&D capital at the beginning of the first year for which R&D expenditure data is available.

R_o = the expenditure on R&D during the first year for which it is available.

g = the average annual logarithmic growth of R&D expenditures over the period for which published R&D data were available, in this case 1978-79 to 2002-03.

δ = the depreciation or obsolescence rate of knowledge, in this case 10 percent.

Bernstein (1989) in his analysis of inter-industry spill-overs in Canada. Moreover, in preliminary analysis of the Australian panel data, we found that using depreciation rates of 8 and 15 percent did not alter the results significantly. Expenditure figures for total R&D were extracted from ABS, Cat. No. 8111.0 whilst figures relating to university R&D were sourced from ABS, Cat No. 8112.0.

According to the ABS (Cat No. 8112), the following definitions apply to the different components of R&D. *Pure basic research* is defined as: "Experimental and theoretical work undertaken primarily to acquire new knowledge without a specific application in view." *Strategic basic research* is defined as research that is: "directed into specified broad areas in the expectation of useful discoveries. It provides the broad base of knowledge for the solution of recognised practical problems." *Applied research* is: "Original work undertaken in order to acquire new knowledge with a specific application in view. It is undertaken either to determine possible uses for the findings of basic research or to determine new methods or ways of achieving some specific and predetermined objectives." Finally, *Experimental research* is: "Systematic work, using existing knowledge gained from research or practical experience, for the purpose of creating new or improved products/processes."

Capital: PIM was also used to calculate the stock of physical capital for each Australian state. It was assumed that the asset life of plant and equipment was 12 years, while buildings would last 40 year[12]. The stock for these two variables was calculated separately and combined to represent the stock of total physical capital (expenditure figures extracted from ABS, Cat. No. 5242.0)

Descriptive statistics for the data used in this analysis are presented in Table 1. In 2002-03, experimental R&D was the largest R&D activity (39% of total R&D), with applied research being 36% of total, and only 10% coming from pure basic research. Almost 80% of the pure basic research originates from the university sector. This contrasts with the bulk (about 86%) of experimental R&D being conducted by the business sector, and roughly one third of applied research coming from the business sector (see ABS Cat. No. 8112).

Table 1. **Descriptive Statistics, 1978-79 to 2002-2003**

	NSW	VIC	QLD	WA	SA	TAS
Gross State Product						
Mean	201296	143810	86726	55507	41113	11507
Std.Dev	48407	29469	24803	16875	7875	1747
Employed						
Mean	2594	1978	1327	628	745	189
Std.Dev	306	205	279	46	137	12
Capital Stock						
Mean	152848	106608	63555	54888	27098	7090
Std.Dev	24235	17797	14468	18435	8761	1156
Total R&D Stock						
Mean	14096	13507	4991	3351	3946	1982
Std.Dev	5858	5345	2315	1932	1396	1373

12 Bureau of Industry Economics (1995).

Table 1. continued

	NSW	VIC	QLD	WA	SA	TAS
University R&D Stock						
Mean	3913	2793	1613	1030	1110	294
Std.Dev	1356	1045	834	491	340	118
Non-University R&D Stock						
Mean	10182	10714	3379	2321	2836	1688
Std.Dev	4507	4302	1484	1441	1063	1285
Pure R&D Stock						
Mean	1552	949	462	305	396	102
Std.Dev	547	336	212	129	96	32
Strategic R&D Stock						
Mean	839	711	399	276	299	78
Std.Dev	284	247	205	152	95	38
Applied R&D Stock						
Mean	1210	962	650	383	351	102
Std.Dev	416	405	371	186	131	47
Experimental						
Mean	311	171	102	66	64	11
Std.Dev	112	62	51	27	24	4
Union Density						
Mean	38	38	37	32	40	46
Std.Dev	9	10	9	9	8	10

Std.Dev = standard deviation

V. RESULTS

The Cobb-Douglas and Translog versions of the Australian states production function were estimated using both pooled Least Squares and Estimated Generalised Least Squares (EGLS), but we present and discuss only the latter set of results. The EGLS estimated here corrects for both cross-section heteroscedasticity, as well as contemporaneous correlation between the cross-sections.[13] According to Beck & Katz (1995), this approach is valid when contemporaneous correlation is present and when the time span is at least three times greater than the number of cross-sections. Both of these conditions are met for out dataset.

The results from estimating the Cobb-Douglas version of the Australian states production function are presented in Table 2. Column 1 presents the basic production function without R&D. The total stock of R&D is added in column 2. As expected, the coefficients on capital stock, labor and total R&D stock were all found to be positive and statistically significant. R&D in Australia is an important input into the production process. Note, however, that the contribution of total R&D is lower than either labor or physical capital. Similar results are reported in column 3, where the total R&D stock variable is replaced by separate variables for the stock of university and non-university R&D. In column 4, university R&D is disaggregated into the four major categories, with the aim of highlighting which types of

[13] All of the regression analysis was performed using Eviews 5.0.

R&D are the most productive. Unionization and deregulation are introduced in the results reported in column 5.

Unfortunately, our preferred specification reported in column 5 has several coefficients with rather surprising results. If these results are to be believed, non-university R&D has a negative effect on output levels, union density has a positive effect and deregulation had no effect at all. All these are opposite of what we expected and are in sharp contrast to the findings of other researchers.

Table 2. Cobb-Douglas Production Function, Australian States, 1978-79 to 2002-2003, Pooled Estimated Generalized Least Squares, (Dependent variable = ln Gross State Product)

Variable	(1)	(2)	(3)	(4)	(5)
Constant	2.151 (81.35)***	2.130 (33.62)***	2.380 (46.65)***	1.845 (41.16)***	1.792 (39.93)***
ln(Capital)	0.520 (76.56)***	0.467 (34.03)***	0.433 (47.08)***	0.483 (63.45)***	0.490 (73.56)***
ln(Labor)	0.484 (64.92)***	0.434 (44.46)***	0.367 (34.11)***	0.348 (23.09)***	0.313 (17.00) ***
ln(Total R&D)	-	0.111 (16.98)***	-	-	-
ln(Non-University R&D)	-	-	0.022 (4.43)***	-0.019 (-2.23)**	-0.020 (-2.49)**
ln(University R&D)	-	-	0.186 (19.00)***	-	-
ln(Pure R&D)	-	-	-	0.244 (13.73)***	0.228 (10.81)***
ln(Applied R&D)	-	-	-	0.308 (10.67)***	0.331 (10.47)***
ln(Experimental R&D)	-	-	-	-0.203 (-9.86)***	-0.188 (-9.28)***
ln(Strategic R&D)	-	-	-	-0.118 (-3.47)***	-0.109 (-3.14)***
Union Density	-	-	-	-	0.002 (3.42)***
Deregulation	-	-	-	-	0.009 (1.03)

*, **, *** statistically significant at the 10%, 5% and 1% levels, respectively. t-statistics using robust standard errors reported in brackets.

Given the restrictive nature of the Cobb-Douglas, we prefer the results from the Translog. Table 3 reports the results from estimating the Translog version of the production function, allowing for important interactions among the variables. Column 1 presents the basic regression results, using Total R&D. Interestingly, none of the Total R&D variables are individually statistically significant[14]. Total R&D is separated into university and non-university R&D in column 2. Column 3 reports the results from disaggregating university R&D into the four types, but without allowing for interaction effects between these types. The

14 However, a Wald test rejects the Null hypothesis that the four Total R&D variables are jointly statistically non-significant (Chi-square = 86.64, prob value = 0.00).

results from the full Translog model are presented in column 4. These are our preferred results.

Table 3. Translog Production Function, Australian States, 1978-79 to 2002-2003, Pooled Estimated Generalized Least Squares, (Dependent variable = ln Gross State Product)

Variable	(1)	(2)	(3)	(4)
Constant	9.07 (17.61)***	10.696 (6.54)***	24.631 (7.71)***	38.642 (7.08)***
ln(Capital)	-1.820 (-8.29)***	-3.289 (-5.44)***	-6.300 (-7.06)***	-6.283 (-6.48)***
ln(Labor)	2.348 (9.72)***	2.794 (10.78)***	3.462 (5.63)***	1.024 (1.11)
ln(Total R&D)	-0.156 (-1.25)	-	-	-
ln(Total R&D) Squared	0.004 (0.52)	-	-	-
ln(Capital)*ln(Total R&D)	0.031 (0.72)	-	-	-
ln(Labor)* ln(Total R&D)	-0.027 (-0.60)	-	-	-
ln(Capital) squared	0.227 (5.69)***	0.659 (9.36)***	0.576 (6.30)***	0.469 (5.36)***
ln(Labor) squared	0.219 (5.68)***	0.363 (5.32)***	0.666 (5.21)***	0.816 (6.30)***
ln(Capital)*ln(Labor)	-0.424 (-6.55)***	-0.897 (-11.24)***	-0.806 (-5.55)***	-0.799 (-5.40)***
ln(Non-University R&D)	-	-2.311 (-6.58)***	-1.913 (-4.11)***	-1.112 (-2.07)**
ln(University R&D)	-	3.701 (7.38)***	-	-
ln(Capital)*ln(University R&D)	-	-1.700 (-8.45)***	-	
ln(Labor)*ln(University R&D)	-	1.403 (6.98)***	-	
ln(University R&D)*ln(Non-University R&D)	-	-0.085 (-2.05)**	-	
ln(Capital)*ln(Non-University R&D)	-	0.968 (6.77)***	1.120 (10.13)***	0.804 (6.28)***
ln(Labor)*ln(Non-University R&D)	-	-0.932 (-5.99)***	-1.483 (-9.62)***	-1.362 (-8.47)***
ln(Non-University R&D) squared	-	-0.060 (-3.85)***	-0.075 (-2.06)**	-0.134 (-4.05)***
ln(University R&D) squared	-	0.385 (6.26)***	-	
ln(Pure R&D)	-	-	1.644 (1.65)	-2.747 (-1.23)
ln(Applied R&D)	-	-	-0.345 (-0.58)	-6.881 (-4.22)***
ln(Experimental R&D)	-	-	2.814 (2.82)***	10.988 (4.03)***
ln(Strategic R&D)	-	-	0.441 (0.52)	2.670 (1.69)*
ln(Pure R&D)* ln(Capital)	-	-	-0.573 (-3.33)***	0.048 (0.20)
ln(Applied R&D)* ln(Capital)	-	-	0.181 (1.35)	0.833 (4.52)***
ln(Strategic R&D)* ln(Capital)	-	-	-0.888 (-5.75)***	-0.947 (-5.36)***
ln(Experimental R&D)* ln(Capital)	-	-	-0.370 (-3.12)***	-0.966 (-5.67)***
ln(Pure R&D)* ln(Labor)	-	-	0.223 (1.18)	0.201 (0.86)
ln(Applied R&D)* ln(Labor)	-	-	-0.671 (-4.18)***	-0.394 (-1.97)*
ln(Strategic R&D)* ln(Labor)	-	-	2.059 (10.86)***	2.006 (9.24)***
ln(Experimental R&D)* ln(Labor)	-	-	-0.128 (-1.04)	-0.560 (-2.66)***
ln(Pure R&D)*ln(Non-University R&D)	-	-	0.143 (0.62)	0.392 (1.71)*
ln(Applied R&D)*ln(Non-University R&D)	-	-	-0.256 (-1.78)*	0.188 (1.27)
ln(Strategic R&D)*ln(Non-University R&D)	-	-	0.200 (2.73)***	0.115 (1.01)
ln(Experimental R&D)*ln(Non-University R&D)	-	-	0.201 (2.05)**	-0.032 (-0.27)
ln(Pure R&D) squared	-	-	0.157 (1.06)	0.338 (1.33)

Table 3. continued

Variable	(1)	(2)	(3)	(4)
ln(Applied R&D) squared	-	-	0.426 (5.25)***	1.061 (2.26)**
ln(Experimental R&D) squared	-	-	0.045 (0.54)	0.721 (2.19)**
ln(Strategic R&D) squared	-	-	-0.594 (-6.14)***	-0.096 (-0.21)
ln(Pure R&D)*ln(Applied R&D)	-	-	-	-0.665 (-1.67)*
ln(Pure R&D)*ln(Strategic R&D)	-	-	-	-0.221 (-0.59)
ln(Pure R&D)*ln(Experiment R&D)	-	-	-	-0.243 (-0.57)
ln(Applied R&D)*ln(Strategic R&D)	-	-	-	-1.159 (-1.28)
ln(Applied R&D)*ln(Experiment R&D)	-	-	-	-0.705 (-2.14)**
ln(Strategic R&D)*ln(Experiment R&D)	-	-	-	0.510 (1.23)
Union Density	-0.003 (-8.79)***	-0.002 (-1.73)*	-0.001 (-1.45)	-0.002 (-2.50)**
Deregulation	0.019 (2.56)**	0.041 (5.97)***	0.042 (6.44)***	0.048 (6.03)***

*, **, *** statistically significant at the 10%, 5% and 1% levels, respectively. t-statistics using robust standard errors reported in brackets.

While some of the results are similar, there are also important differences between the results from the Cobb-Douglas and the Translog specifications. The key elasticities are compared in Table 4 (the elasticities for the Translog are calculated at the sample means). In both cases, the elasticity of capital is greater than it is for labor. In both cases, pure basic and applied R&D have a positive impact on output, while strategic basic and experimental R&D have an adverse effect on output. The Translog indicates a lower elasticity associated with Pure and Applied R&D. The key differences are the elasticity of non-university R&D and university R&D. The Cobb-Douglas specification suggests a small negative non-university R&D elasticity, while the Translog suggests a more plausible positive elasticity. Importantly, the Cobb-Douglas specification suggests a positive university R&D elasticity, while the Translog suggests a small negative elasticity.[15]

Table 4. Australian Gross State Product (GSP) Elasticities, Various Specifications

Elasticity	Cobb-Douglas (Table 2, column 5)	Translog (Table 3, column 4)
Elasticity of GSP w.r.t Capital	0.49 [5411.29]***	0.62 [2157.89]***
Elasticity of GSP w.r.t Labor	0.31 [288.91]***	0.37 [287.97]***
Elasticity of GSP w.r.t Non-University R&D	-0.02 [6.18]**	0.11 [197.82]***
Elasticity of GSP w.r.t University R&D	0.26 [324.85]***	-0.03 [307.90]***
Elasticity of GSP w.r.t University Pure R&D	0.23 [116.85]***	0.15 [44.48]***
Elasticity of GSP w.r.t University Applied R&D	0.33 [109.57]***	0.10 [196.91]***
Elasticity of GSP w.r.t University Strategic R&D	-0.11 [9.86]***	-0.17 [203.22]***
Elasticity of GSP w.r.t University Experimental R&D	-0.19 [-86.16]***	-0.10 [75.61]***

Figures in squared brackets are Wald tests, testing whether the associated elasticity is equal to zero. **, *** statistically significant at the 5% and 1% levels, respectively.

[15] That is, if pure, applied, strategic and experimental R&D all rise by 1 percent, Australian state output will, on average, decline by 0.03 percent.

While it is of small magnitude, the negative elasticity of output with respect to university R&D is of some concern, especially from a policy perspective. It is clear, however, that this is driven by the negative elasticities associated with strategic and experimental R&D. The results on both applied and experimental are not surprising. We expect that R&D effort devoted towards applications of basic research should have positive payoffs. Experimental may be expected to go either way, although those who initiate it would expect to get returns from it. However, pure basic R&D is not directed towards long term benefits and is undertaken for the sake of advancing knowledge. Hence, it is heartening to find that the advancement of knowledge *per se* has a positive effect on the economy. It was surprising to us that strategic basic R&D had a robust negative effect (regardless of the specification used). The reasons for this are beyond the scope of this paper, but do warrant a full examination.

Over the period studied, unionization had a consistently negative impact on state output, but the effect was small.[16] The results presented here also indicate that deregulation of the financial sector had a favorable impact on state output. State output is about 5 percent higher in the post deregulation period.

Additional insights into the responsiveness of output with respect to university R&D can be gleamed from the interaction effects. These are summarized in Table 5.

Table 5. Substitution and Complementarity in Australian State Output, 1979-2003

	Capital (1)	Labor (2)	Non-University R&D (3)	Pure (4)	Applied (5)	Strategic (6)	Experimental (7)
Capital	Comp	Sub	Comp	Neutral	Comp	Sub	Sub
Labor	Sub	Comp	Sub	Neutral	Sub	Comp	Sub
Non-University R&D	Comp	Sub	Sub	Comp	Neutral	Neutral	Neutral
Pure	Neutral	Neutral	Comp	Neutral	Sub	Neutral	Neutral
Applied	Comp	Sub	Neutral	Sub	Comp	Neutral	Sub
Strategic	Sub	Comp	Neutral	Neutral	Sub	Neutral	Neutral
Experimental	Sub	Sub	Neutral	Neutral	Sub	Neutral	Comp

Sub = substitute. Comp = complement. All Sub and Comp associations are statistically significant.

There are three sets of important university R&D interactions: with factor inputs; with non-university R&D; and among the different types of R&D.

Impact on Marginal Productivity of Labor and Capital

As expected, capital and labor are substitutes. Non-university R&D is a complement with capital and a substitute with labor. Pure R&D has no effect at all on the marginal productivity of either capital or labor. Applied R&D is complementary with capital but is a substitute with labor. Hence, its effect is the same as non-university R&D. Strategic R&D has the opposite effect. Experimental R&D is a substitute with both factors.

---

16 We do not address, however, other effects of unions, such as provision of benefits to employees, safety in workplaces etc. These issues need to be addressed in future research. Rogers (2000) also briefly address the issue of unions. Using probit regressions, union presence was represented by a dummy variable. The results indicated that firms with unions had a higher probability of reporting to be innovative.

Impact on Marginal Productivity of Non-University R&D

Pure is the only type of university R&D that has an impact on non-university R&D. Pure R&D is complementary with non-university R&D. Increasing pure university R&D has a positive effect on the contribution of non-university R&D to output. Since the other types of university R&D are neutral, we conclude that university R&D has a positive spillover effect on non-university R&D.

Non-Linearities

The squared input terms capture the existence of non-linear effects. The contributions of capital and labor are greater at higher levels of input usage. However, non-university R&D's contribution declines at higher levels of non-university R&D. Both strategic and pure research are neutral, while applied and experimental are both complementary. That is, the contributions of applied R&D to production are higher at higher levels of applied R&D.

Interactions between University R&D

The only effect pure research has is on applied R&D, where it is, surprisingly, a substitute in the production process. Applied R&D is a substitute also with experimental R&D. Experimental and strategic are neutral with respect to each other's impact on production.

Given these results, it is beneficial to briefly examine the relative share of each of the four types of university R&D. Data for selected years is presented in Table 6. Over the 1979-2003 time period, the relative share of pure R&D undertaken by the university section has decreased. This decrease will have had a negative effect on output levels. This has been offset by the corresponding increase in the share of applied R&D. The increase in applied R&D will have had a positive effect on output levels. In light of the results presented here, it is tempting to conclude that the university sector should shift the focus away from experimental and particularly strategic research and towards pure and applied research areas, which have a positive impact on Australia's overall productivity. However, it is possible that strategic and experimental R&D may have other long-run benefits to production. These need to be considered before firmer inferences can be drawn to guide policy.[17]

Table 6. Relative Share of University R&D Stocks, by type, selected years

Types of Research	1978-79 %	1988-89 %	2002-03
Pure	35	35	30
Strategic	25	25	26
Applied	34	34	38
Experimental	6	6	6

Table 7 uses the parameter estimates from Table 3 and the growth accounting framework to identify the sources of regional growth. The first row reports the average annual rate of output growth, and the other rows report the contributions of the various inputs and sources of

[17] For example, it is possible that strategic R&D contributes to applied R&D and, hence, has a positive indirect effect on output.

R&D. For example, NSW GSP grew at an average annual rate of 3.32%. Of this, 1.34% (or 40 percent of NSW's growth) can be attributed to growth in physical capital, 0.60% to growth in labor, 0.72% to growth in non-university R&D and 1.20% to growth in pure and applied university R&D. In most cases, the contributions of non-university R&D were larger than those of labor. In the case of Tasmania, non-university R&D was more important than even physical capital formation. Interestingly, with the exception of Tasmania, the contributions of university pure and applied R&D were higher than those from the non-university sector.

Table 7. Contributions of Factor Inputs and R&D to Australian State Output Growth, 1979-2003 (all figures are percentages)

	NSW	VIC	QLD	WA	SA	TAS
Output	3.32	2.83	3.96	4.40	2.70	2.10
Capital	1.34	1.42	1.91	2.95	2.74	1.42
Labor	0.60	0.52	1.11	0.39	0.95	0.30
Non-University R&D	0.72	0.64	0.68	1.00	0.60	1.48
Total University R&D	-0.12	-0.10	-0.20	-0.27	-0.16	-0.32
- Applied and Pure R&D	1.20	1.33	1.73	1.63	0.98	1.33
- Strategic and Experimental R&D	-1.32	-1.43	-1.93	-1.90	-1.14	-1.65

VI. Concluding Comments

The importance of R&D to the production process has long been a source of considerable discussion and evaluation. Numerous international studies have highlighted its contribution to productivity and have concluded that it is an important source of economic growth. A review of Australian literature in this field reveals that relatively little has been written on the topic outside of agriculture and manufacturing industries. Furthermore, to date, the impact of higher education R&D on Australian productivity has been an area which has been largely ignored.

This paper has examined the impact of R&D on Australian state production, and in particular that conducted by the higher education sector, through Cobb-Douglas and Translog production functions. The results confirm that, in addition to capital and labor, R&D is an important and vital input in the production process.

Analysis of the four types of R&D carried out by higher education organizations revealed that both pure and applied R&D have a statistically significant positive impact on Australian regional economic performance. However, strategic and experimental R&D have had an adverse effect, although that they may have other payoffs that remain unexplored. Nevertheless, over the last two decades the majority of research undertaken by the higher education sector in Australia has been in the form of pure and applied research, and these have contributed to regional development.

The results are robust and show that, on average, experimental and strategic basic R&D have negative returns, while pure basic and applied research have positive returns. This is, of course, an average effect, and there could be many projects that fall in the first two categories that have positive payoffs. However, it is clear from the results that Australian regional production benefits more from universities' pure basic research than from strategic basic research. This is an interesting finding that warrants further investigation. An interesting extension would be to compare strategic basic research conducted in the non-university sector to that conducted in the university sector, with the view of exploring the productivity differences between the two.

It is important to note that the production function framework we have used informs on the marginal productivity of factors and how university R&D affects this. This is the direct effect. Additionally, there may be important indirect effects. For example, university R&D may stimulate non-university R&D and may also have important factor accumulation effects. These need to be modeled and explored before firm conclusions can be reached regarding the total effect of university R&D on output.

The focus in this paper has been on total R&D. It would be very interesting to extend the analysis to compare the productivity effects of R&D expenditure across different fields, such as engineering, sciences, information systems and medicine. Furthermore, it would be interesting to compare the impact of different objectives. For example, most of the R&D is directed towards economic development, while other R&D is directed towards defense, environment, health and general societal objectives. Comparing the economic impact of these types of R&D would provide a more complete analysis of the effects of university R&D.

REFERENCES

Acs, Z., Audretsch, D. and Feldman, P. (1992) 'Real Effects of Academic Research', *American Economic Review* vol. 82, no. 1, pp. 363-67.

- (1994a) 'R&D Spillovers and Innovative Activity', *Managerial and Decision Economics*, vol. 15, no. 2, pp.131-138.

- (1994b) 'R&D Spillovers and Receipt Firm Size', *The Review of Economics and Statistics*, vol. 76, no.2, pp. 336-40.

Addison, J. and C. Belfield. (2004). 'Union Voice', *Journal of Labor Research*, vol. 25, no. 4, pp. 563-96.

Alexander, M. and Green, R. (1992) 'Workplace Productivity and Joint Consultation', *Australian Bulletin of Labour,* vol. 18, no.2, pp.95-118.

Anselin, L., Varga, A. and Acs, Z. (1997) 'Local Geographic Spillovers Between University Research and High Technology Innovations', *Journal of Urban Economics*, vol. 42, no. 3, pp. 422-448.

- (2000) 'Geographic Spillovers and University Research: A spatial econometric perspective', *Growth and Change*, vol. 31, no. 4, pp.501-15.

Australian Bureau of Statistics (various dates) *Research and Experimental Development, Higher Education Organisations, Australia*, Cat. No, 8110.0, AGPS, Canberra.

Australian Bureau of Statistics, *All Sector Summary, Australia, 2002-03*, Cat. No. 8112.0.

Batterham, R. (2000) *The Chance to Change: Final report by the Chief Scientist*, Department of Industry, Science and Resources, Canberra.

Beck, N and Katz, J. (1995) 'What to do (and Not to Do) with Time-Series Cross-Section Data', *American Political Science Review*, vol. 89, no. 3, pp. 634-647.

Bernstein, J. (1988) 'Costs of Production, Intra- and Interindustry R&D Spillover: Canadian evidence' *The Canadian Journal of Economics,* vol. 21, no.2, pp.324-47.

- (1989) 'The Structure of Canadian Inter-industry R&D Spillovers, and the Rates of Return to R&D' *The Journal of Industrial Economics,* vol. 37, no. 3, pp.315-328.

Bernstein, J. and Nadiri, M. (1988) 'Interindustry R&D Spillovers, Rates of Return and Production in High-Tech Industries' *American Economic Review,* vol. 78, no. 2, pp. 429-434.

Brambles, T. (2001) 'Australian Union Strategies Since 1945', *Labour and Industry,* vol.11, no. 3, pp. 1-26.

Briggs, C. (2001) 'Australian Exceptionalism: The role of trade unions in the emergence of Enterprise Bargaining', *Journal of Industrial Relations,* vol. 43, no. 1, pp. 27-43.

Cabalu, H., Kenyon, P. and Koshy, P. (2000) *Of Dollars and Cents: Valuing the Economic Contribution of Universities to the Australian Economy,* Commissioned Report (Business/Higher Education Round Table): no.5.

Coe, D. and Helpman, E. (1995) 'International R&D Spillovers', *European Economic Review,* vol. 39, no. 5, pp. 859-887.

Coe, D. and Moghadam, R. (1993) 'Capital and trade as engines of growth in France: An application of Johansen's cointegration methodology', *IMF Staff Papers,* vol. 40, no. 3, pp. 542-66.

Cuneo, P. and Mairesse, J. (1984) 'Productivity and R&D at the Firm Level in French Manufacturing', in Z. Griliches (ed) *R&D Patents and Productivity,* University of Chicago Press, Chicago.

Commonwealth of Australia (2001) *Backing Australia's Ability: An Innovation Action Plan for the Future,* Commonwealth of Australia, Canberra.

Doucouliagos, C. and Hone, P. (2000) 'Deregulation and Subequilibrium in the Australian Dairy Processing Industry', *The Economic Record,* vol. 76, no. 233, pp. 152-62.

Doucouliagos C. & Laroche P. (2003), 'What do unions do to productivity? A meta-analysis', *Industrial Relations,* vol. 42, no. 4, pp. 650-691.

Dowrick, S. (1994) '*The Role of R&D in Growth: A Survey of the New Growth Theory and Evidence*', Industry Commission.

Drago, R. and Wooden, M. (1992) 'The Australian Workplace Industrial Relations Survey and Workplace Performance', *The Australian Bulletin of Labour,* vol. 18, no. 2, pp. 142-167.

Drago, R., Wooden, M. and Sloan, J. (1992) *Productive relations? Australian Industrial Relations and Workplace performance,* Allen & Unwin, Sydney.

Fenney, S. and Rogers, M. (2001) 'Innovation and Performance: Benchmarking Australian firms', *Melbourne Institute of Applied Economics and Social Research,* Working Paper No. 7/01.

Freeman, R. and Medoff, J. (1984) *What do unions do?,* Basic Books, New York.

Goto, A. and Suzuki, K. (1989) 'R&D Capital, Rate of Return on R&D Investment and Spillover of R&D in Japanese Manufacturing Industries' *The Review of Economics and Statistics,* vol. 71, no. 4, pp. 555-564.

Griliches, Z. (1964) 'Research Expenditure, Education and the Aggregate Agricultural Production Function', *American Economic Review,* vol. 54, no. 6, pp.961-974.

- (1973) 'Research Expenditure and Growth Accounting', in B.R. Williams (ed) *Science and Technology in Economic Growth,* John Wiley & Sons, Halsted Press, New York.

- (1979) 'Issues in Assessing the Contribution of Research and Development to Productivity Growth', *Bell Journal of Economics,* vol. 10, no. 1, no. 92-116.

- (1990) 'Patent Statistics as Economic Indicators: A Survey', *Journal of Economic Literature*, vol. 28, no. 4, pp.1661-1707.

- (1992) 'The Search for R&D Spillovers', *The Scandinavian Journal of Economics,* (Supplement), vol. 94, no. 0, pp. S29-47.

- (1994) 'Productivity, R&D and Data Constraints', *American Economic Review*, vol.84, no. 1, 1-23.

Griliches, Z. and Mairesse, J. (1984) 'Productivity and R&D at the Firm Level', chapter 17 in Z. Griliches (Ed) *R&D, Patents and Productivity,* University of Chicago Press, Chicago.

Groenewold, N. (2003), 'Financial Deregulation and the Relationship between the Economy and the Share Market in Australia' *Australian Economic Papers*, vol. 42, no. 4, pp. 454-77.

Gwartney, J. and R. Lawson. (2003) *Economic Freedom of the World Annual Report.* The Fraser Institute, Vancouver, BC.

Health and Medical Research Strategic Review (1999) *The Virtuous Cycle: Working together for health and medical research*, Department of Health & Aged Care, Canberra.

Hirsch, B. 2004. 'What Do Unions Do for Economic Performance?' *Journal of Labor Research*, vol. 25, no. 3, pp. 415-55.

Industry Commission (1995) *Research and Development*, Report No 44, May 15, vol. 3, Government Publishing Services, Canberra.

Jaffe, A. (1986) 'Technological Opportunity and Spillovers of R&D: Evidence from firms' patents, profits and market value', *American Economic Review*, vol. 76, no. 5, pp. 984-1001.

- (1989) 'Real Effects of Academic Research' *American Economic Review,* vol.79, no. 5, pp.957-970.

Kemp, D. (1999) *Knowledge and Innovation: A policy statement on research and research training,* Department of Education, Training and Youth Affairs, Canberra.

Lie, F., Brooks, R. and Faff, R. (2000) 'Modelling the Equity Beta Risk of Australian Financial Sector Companies', *Australian Economic Papers*, vol. 39, no.3. pp.301-311.

Link, A. & Rees, J. (1990) 'Firm Size, University-Based Research and the Returns to R&D' *Small Business Economics*, vol. 2, no. 1, pp.25-32.

Mairesse, J. and Mohnen, P. (1994) 'R&D and Productivity Growth: What have we learned from econometric studies?' In *Proceedings of the EUNETIC (European Network on the Economics of Technological and Institutional Change) Conference on Evolutionary Economics of Technological Change: Assessment of Results for the New Frontiers* Strasbourg: BETA Strasbourg Communaute Urbaine.

Mairesse, J. and Sassenou, M. (1991) 'R&D and Productivity: A survey of econometric studies at the firm level', *Science-Technology Industry Review*, vol. 8, pp. 9-43.

Mansfield, E. (1965) 'Rates of Return from Industrial R&D', *American Economic Review,* vol. 70, pp. 863-873.

- (1972) *Contribution of Research and Development to Economic Growth in the United States,* Papers and Proceedings of a Colloquium on Research and Development and Economic Growth Productivity, National Science Foundation, Washington D.C.

- (1984) 'Comments on Using Linked Patent and R&D Data to Measure Interindustry Technology Flows', in Griliches, Z (ed) *R&D Patents and Productivity*, University of Chicago Press, Chicago.

Minasian, J.R. (1962) 'The Economics of Research and Development', in R.R. Nelson (ed) *The Rate of and Direction of Inventive Activity: Economics and Social Factors*, NBER Special Conference vol. 13, Princeton University Press, Princeton.

Mohnen, P. (1990) *R&D and Productivity Growth: A survey of the literature,* CERPE – Cahier de Recherche no 57, Centre de Recherche sur les Politiques Economiques, Universite du Quebec a Montreal.

- (1996) 'R&D Externalities and Productivity Growth', *Science-Technology Industry Review*, vol. 18, pp. 39-66.

Mohnen, P and Lepine, N. (1988) *Payments for Technology as a Factor of Production*, Cahier n. 8810, Departement des Sciences Economiques, Universite du Quebec a Montreal.

- (1991) 'R&D, R&D Spillovers and Payments for Technology: Canadian Evidence', *Structural Change and Economic Dynamics,* vol. 2, no. 1, pp. 213-28.

Mohnen, P., Nadiri, M. and Prucha, I. (1986) 'R&D, Production Structure and Rates of Return in US, Japanese and German Manufacturing Sectors: A non-separable dynamic factor demand model', *European Economic Review*, vol. 30, no. 4, pp. 749-772

Nadiri, M.I. (1980) 'Sectoral Productivity Slowdown', *American Economic Review*, vol. 70, no. 2, pp. 349-52.

- (1993) 'Innovations and Technological Spillovers', *NBER Working Paper no. 4423*, National Bureau of Economic Research, Cambridge.

Nelson, R. (1986) 'R&D, Innovation and Public Policy', *American Economic Review,* vol. 76, pp.186-89.

OECD (1980) *The Measurement of Scientific Technical Activities: Proposed Standard Practice for Surveys of Research and Experimental Development - Frascati Manual*

Olley, S. and Pakes, A. (1996) 'The Dynamic of Productivity in the Telecommunications Equipment Industry', *Econometrica*, vol. 64, no. 6, pp. 1263-1297.

Phipps, A. and Sheen, J. (1992) 'Unionisation, Industrial Relations and Labour Productivity Growth in Australia: A pooled time-series/cross section analysis of TFP growth', *Labour of Economics and Productivity*, vol. 6, pp.54-77.

Rogers (2000) 'Understanding Innovative Firms: An Empirical Analysis of the GAPS' *Melbourne Institute of Applied and Social Research,* Working Paper No. 8/00.

Rogers, M. (2001) 'The International Diffusion of Technology: Technological catch-up and economic growth', in Bloch, H. and Kenyon, P. (eds), *Creating an Internationally Competitive Economy,* Palgrave, New York.

- (2002) 'Firm Performance and Investment in R&D and Intellectual Property', *Melbourne Institute of Applied and Social Research,* Working Paper No. 15/02.

Sarafoglou, N. and Hayes, K. (1996) 'University Productivity in Sweden: A demonstration and explanatory analysis for economics and business programs', *The Annals of Regional Science*, vol. 30, pp. 285-304.

Scherer, F. (1983) 'The Propensity to Patent', *International Journal of Industrial Organization*, vol. 1, no. 1, pp. 107-128.

Sterlacchini, A. (1989) 'R&D, Innovations, and Total Factor Productivity Growth in British Manufacturing', *Applied Economics*, vol. 21, no. 1, pp. 1549-62.

Terleckyj, N. (1958) 'Factors Underlying Productivity: Some empirical observations', *Journal of the American Statistics Association*, vol. 53, pp. 593.

(1974) *Effects of R&D on the Productivity of Growth of Industries: An exploratory study*, National Planning Association, Washington.

Turnbull, P. (2003). 'What Do Unions Do Now?', *Journal of Labor Research*, vol. 24, no. 3, pp.491-527.

In: Economics of Technological Change
Editor: Thomas L. Beane, pp. 89-106
ISBN 1-60021-237-9
© 2007 Nova Science Publishers, Inc.

Chapter 5

COMPARING NONPROFIT VERSUS FOR-PROFIT INNOVATION IN THE U.S. AGRICULTURAL BIOTECHNOLOGY INDUSTRY

H. Phoebe Chan*
Economics Department
University of Michigan-Flint

Abstract

In the 1980s, the United States established patent protection for biotechnology. Following this, patent applications for agricultural inventions have grown significantly. Nonprofits produced many significant agricultural innovations in the past. Now for-profits are producing an increasing share of U.S. agricultural innovations. Two measures of innovation are considered: the number of new agbiotech patents and plant variety protection (PVP) certificates. This research explores the similarities and differences in the roles nonprofits and for-profits have played in agricultural innovation in the past 40 years. The data reveals that for-profit organizations combined own greater numbers of patents and PVP certificates than nonprofits combined. However, the average nonprofit organization owns more patents than the average for-profit organization. The average nonprofit is also likely to hold more patents in DNA technologies than the average for-profit. Finally, nonprofits overall have created more plant varieties than for-profits in particular crop types such as oats, potato, peanut, and rice.

1. Introduction

In the past forty years, several significant events have changed the nature of innovation in agriculture. They have occurred both in the type of technology used in innovation and the intellectual property protection available for the innovations. As a result, nonprofit organizations and for-profit firms have played different and changing roles in industry innovation. In this chapter, innovation is measured as the number of new agbiotech patents and new

*E-mail address: chanp@umflint.edu

plant varieties protection certificates. The types and amount of innovation in the U.S. agricultural industry are compared using these two measures between the years 1971 and 2005 for nonprofits versus for-profits.

In 1970, the Plant Variety Protection (PVP) Act was passed. It provided a new form of intellectual property protection for sexually-reproduced plant varieties [1]. According to Butler and Marion (1985), for-profit innovation increased but not uniformly across crop varieties. They also document that some nonprofit organizations, namely agricultural experiment stations, felt pressure to decrease the creation of new plant varieties and re-focus their energies on more basic research. At this time, most new plant varieties were created using traditional breeding methods. That is, firms would use cross-pollinization to create new plant hybrids from other plant varieties identified as having desirable characteristics.

Beginning in the 1980s, biotechnology enabled scientists to link desirable characteristics found in an organism to particular genes. Using biotechnology, scientists could isolate desirable genes and insert them into an existing plant variety to create a new plant variety. With biotechnology, the organism from which the genes are donated need not only be from plants but may also be from animals or bacteria. Scientists now have even greater opportunities to create new plant varieties. First generation products include plants with herbicide tolerance, insect resistance, and increased yields.

At approximately the same time biotechnology began being utilized in agricultural innovation, the U.S. began allowing firms to protect both research related to biotechnology and new plant varieties with utility patents [2]. This increased incentives for for-profit firms' to participate in plant variety research. Since many developing countries depend on a thriving agricultural industry, many scholars expressed concern over the welfare effects of for-profits' increased involvement in agricultural innovation [3]. (Barton, 1998; Graff, Rausser and Small, 2003; King and Schimmelpfennig, 2005)

This research provides a descriptive analysis of the U.S. agricultural biotechnology industry, comparing the creation of biotechnology inventions and new plant varieties by U.S. nonprofit organizations to for-profit firms. Specifically, the number of U.S. agbiotech patents issued to U.S. nonprofit organizations and for-profit firms are examined between the years 1976-2000. Plant variety creation is examined using the number of new U.S. plant variety protection (PVP) certificates applied for between the years 1971-2005. For-profit firms are defined as U.S. companies while nonprofit organizations include U.S. government agencies, U.S. universities, and U.S. non-governmental organizations. King and Heisey's (2004) agricultural biotechnology intellectual property (ABIP) database is used to identify U.S. agbiotech patents. The U.S. Department of Agriculture's Agricultural Marketing Service PVP certificates status database is used to identify plant variety protection certificates.

Previous papers have discussed the roles of nonprofits versus for-profits in innovation. Butler and Marion (1985) note that prior to the establishment of PVP certificates, most plant varieties were created by nonprofits, specifically agricultural experiment stations and public

[1]Specifically, sexually-reproduced plant varieties or plants produced from seeds. As well, plant patents have been available for asexually reproduced plants since 1930.

[2]Specifically, the U.S. Supreme Court case of Diamond v. Chakrabarty in 1980 paved the path towards permitting patents for genes and living organisms.

[3]For-profits could decrease future innovation. As well, compared to nonprofit creation of new plant varieties, for-profits have fewer incentives to share new technologies with developing countries.

institutions, with the exception of corn and sorghum. Meanwhile many (e.g., Mowery, 2004; Zilberman, Yarkin, and Heiman, 2001; and Graff, 1997) note that the biotechnology industry[4] itself also originated from nonprofit institutions. Zilberman, Yarkin, and Heiman (2001) declare that "in spite of the privatization of knowledge, universities are still the major source of biotechnology innovations."

Nonprofit innovation may play complementary and/or substitute roles to for-profit innovation. For instance, nonprofits have launched close to 100 start-ups some of which were later acquired by for-profits. Further, both the Camarosa strawberry and baby carrots were a result of successful technology transfer between nonprofits and for-profits. (Parker, Castillo, and Zilberman, 2001) However, Naseem and Oehmke (2004) also discuss the patent races that exist between nonprofits and for-profits in rice genomics. This research examines whether nonprofit innovation focuses on the same types of technologies and plant species as for-profit innovation.

According to the data, for-profit firms generally create greater numbers of patented technologies. However, the average patent owned by nonprofits versus the average patent owned by for-profits does not appear to differ in the number of claims or the number of technology types spanned. Also, the average nonprofit organization will produce a greater number of patented technologies than the average for-profit firm. Further, a greater share of nonprofit organizations specialize in DNA-related technologies than for-profit organizations. PVP certificates data reveals that nonprofits have created the majority of new varieties in several plant species such as rice, where varieties are considered to be of significant importance for developing countries. However, they have also created new varieties in many of the same plant species as for-profits. While this chapter describes similarities and differences in the innovative roles that nonprofits and for-profits have played in agbiotech in the last 40 years, this chapter leaves open the discussion of the causal determinants for these events and normative interpretations of the results for future researchers. Finally, this chapter focuses on the creation of new biotechnologies and plant varieties, not the further development or market value of the inventions or new varieties[5].

2. Agbiotech Intellectual Property Protection

This research focuses on two forms of agbiotech intellectual property protection available in the U.S.: utility patents and plant variety protection (PVP) certificates. Utility patents provide a temporary monopoly to the patent owner. That is, the firm can prevent others from making, using, selling, or importing the invention without the explicit consent of the owner. Utility patents can be used to protect several types of innovations related to biotechnology. For example, Graff, Rausser, and Small (2003) divide utility patents into three categories: technologies for plant transformation; gene sequences and genetically coded traits and enhancements; and elite germplasm.

Generally, for-profit firms have high incentives to protect inventions of marketable value with a utility patent. By protecting their inventions with a utility patent, they may license to

[4]The biotechnology industry includes technology related to pharmaceuticals, animals, and plants.

[5]That is, not all patents may be incorporated into a product sold on the market. As well, not all new plant varieties are marketable.

another or further develop and market the invention on their own, allowing them to recover the costs of producing the invention and make profits. The number of utility patents owned by for-profits increased substantially throughout the early 1990s. (ABIP, 2004)

In contrast, the philosophies of nonprofit organizations have changed over time regarding patents. For instance, in the early 1900s, very few universities had patent policies. But by the late 1940s, virtually all major universities had patent policies. Over the last century, universities have steadily increased their ownership of patents. (Mowery, 2004) In 1980, the Bayh-Dole Act was passed, creating a uniform policy that gave universities and businesses patent ownership for research produced under federal contracts[6]. This enabled nonprofits to form exclusive licenses with for-profit firms designed to shorten the time span between inventing and marketing a new product. Since then, nonprofits have increasingly sought patent protection for and engaged in licensing of their inventions. (Parker, Castillo, and Zilberman, 2001) However, Mowery (2004) also states the Bayh-Dole Act was only the most recent of many policies that were enacted in the past half-century to induce increases in patent and licensing activities by universities.

Zilberman, Yarkin, and Heiman (2001) note that nonprofits are more likely to engage in creating innovations focused on basic research (technologies for plant transformation and gene sequences). Basic research is more uncertain to be profitable and involve considerably larger amounts of time and money to make them commercially viable. It follows that nonprofits may in fact be instrumental to producing innovations responsive to developing countries' agricultural needs, where for-profits are less likely to obtain large profits or have the ability to recover their high research and development costs.

A second and alternative type of intellectual property protection is PVP certificates. PVP certificates provide intellectual property protection similar to but weaker than the utility patent. In comparison to utility patent protection, plant variety protection (PVP) certificates may be used to protect only one particular type of agbiotech invention: plant varieties. In fact, PVP certificates were the only available form of intellectual property protection for sexually-reproduced plant varieties[7] before the 1980s. Naseem, Oehmke, and Schimmelpfennig (2005) suggest that for cotton varieties, PVP certificates have provided an effective stimulus for innovation. However, Janis and Kesan (2002) suggest that PVP certificates have become less popular due to the lengthening approval process.

3. Data

Agbiotech data is obtained from King and Heisey's (2004) agricultural biotechnology (ABIP) database. Prior to the year 2000, the U.S. did not publish utility patent applications. Therefore, data is unavailable on failed patent applications. King and Heisey's ABIP database provides data on patents issued between the years 1976 and 2000. King and Heisey divide biotechnology into nine main technology classes which includes technologies related

[6]The government maintains "march-in" rights to force compulsory licenses if it feels that public availability of the invention is being impeded. It is also allowed access to the invention through nonexclusive royalty-free licenses.

[7]Firms could protect asexually-produced (include budding, cutting, division, grafting, and layering) plants with plant patents. (Butler and Marion, 1985)

to both plants and animals. In an effort to focus primarily on plant technologies [8], the sample includes patents related to only six main technology classes: plant technologies; metabolic pathways and biological processes in plants; protection, nutrition, and biological control of plants; genetic transformation; metabolic and biological processes, DNA-scale; and genomics. Below is a brief description the six broad technology classes included in King and Heisey's ABIP database below.

Plant technologies Patents relate to improved agronomic traits such as insect resistance, herbicide tolerance, and improved yields for specific plant species. Plant varieties (also protectable by PVP certificates) are included in this technology class.

Metabolic pathways and biological processes in plants Plant technology related to plant processes on a macro-level scale. Inventions are less applicable to all plants but rather to specific crop types (e.g., corn, coffee, bananas).

Protection, nutrition, and biological control of plants Patents for fertilizers and on methods for increasing the nutritional content of plants.

Genetic transformation Patents on methods and techniques used to aid and implement genomics research. This includes gene markers and transformation platforms. Patents are generally applicable to all plants across multiple species.

Metabolic pathways and biological processes in plants, DNA-scale Patents on technology related to plant processes at the DNA level. Inventions are more applicable to plants across multiple species.

Genomics Patents on sequencing, identifying, isolating, and characterizing gene sequences. Genomics research can be used across multiple crop types.

A single patent may be classified across multiple categories. The dominant for-profit firms across all major technology classes are generally Pioneer Hibred and Monsanto [9]. The dominant nonprofit players are the University of California and the United States Department of Agriculture (USDA). The category under which the most number of patents is classified is genetic transformation. For all categories, the top patent holders are always for-profit firms. Table 1 lists the number of patents in each category and divides them into the share held by nonprofits and the share held by for-profits.

King and Heisey (2004) also divide patent owners into six categories: non-U.S. commercial firms, U.S. commercial firms, non-U.S. nonprofit agencies and universities, U.S. nonprofit agencies and universities, non-U.S. government agencies, and U.S. government agencies. Only patents awarded to U.S. commercial firms, U.S. nonprofit agencies and universities, and U.S. government agencies are included in the analysis. The former category is designated as "for-profit" firms and the latter two categories are designated as "nonprofit" organizations. There are a total of 5207 patents; 89 are jointly owned by both U.S. nonprofits and for-profits. These are removed from the sample; the remaining sample contains 1817 patents owned by nonprofit organizations and 3301 patents owned by for-profit firms.

[8]However, inevitably some patents related to animal biotechnology are still included in the dataset.

[9]King and Schimmelpfennig (2005) discuss firm strategies in the diversification of their patent portfolios.

Table 1. Number of patents by category 1976-2000. Lists the number of patents in each technology classification summed over the years 1976-2000. See the text for explanations of the categories. A single patent may belong to multiple categories.

Category	Number Patents	Number Owned by Nonprofits	Number Owned by For-profits
Plant Tech	2032	470	1562
Met Paths and Biol Proc	744	234	510
Nutrition	1110	502	608
Gene Transf	2771	964	1807
Met Path & Biol Proc DNA	817	350	467
Genomics	188	86	102
Num Obs			5118

Source: King and Heisey ABIP (2004)

Plant variety protection (PVP) certificates data originate from the U.S. PVP Office (2006). The PVP Office lists all PVP applications between the years 1971-2005 [10]. Unlike King and Heisey's ABIP, the PVP Office includes data on both applications and issued certificates. Specifically, they provide the following information on all applications applied for between the years 1971 and 2005: application date, the status of the application, the issue date (when applicable), the crop type, the variety name, and the applicant's name. Data is gathered on all available plant species. Applicants are classified as nonprofit organizations or for-profit firms. The sample is limited to U.S. nonprofit organizations and U.S. firms. There are a total of 6512 PVP certificate applications.

4. Utility Patents

Figure 1 illustrates all patents issued between the years 1971 and 2000 by their application years. According to the figure, the number of successful patent applications generally increases until 1995 and then declines. The decline after 1995 is primarily due to data truncation; that is, the numbers of patents issued after 2000 (but applied for before 1999) are not included in the database [11].

According to Figure 1, the number of issued patents owned by for-profits exceeds the number owned by nonprofits in every application year. It is difficult to ascertain any direct impacts of the passing of the Bayh-Dole Act from Figure 1; i.e., there does not appear to be a stark increase in the number of successful agbiotech patents for nonprofits immediately following the passing of the Bayh-Dole Act in 1980. In fact, the greatest increases appear one decade after Bayh-Dole. Since for-profits experience a similar increase, the increase in nonprofit patent activity likely results more from the advent of biotechnology and the ability to patent gene and plant varieties than the Bayh-Dole Act.

[10] Data was gathered as of March 3, 2006.

[11] Specifically, if the data is examined by the years the patents were issued rather than the years they were applied for, there is no evidence of a decline in the number of issued patents.

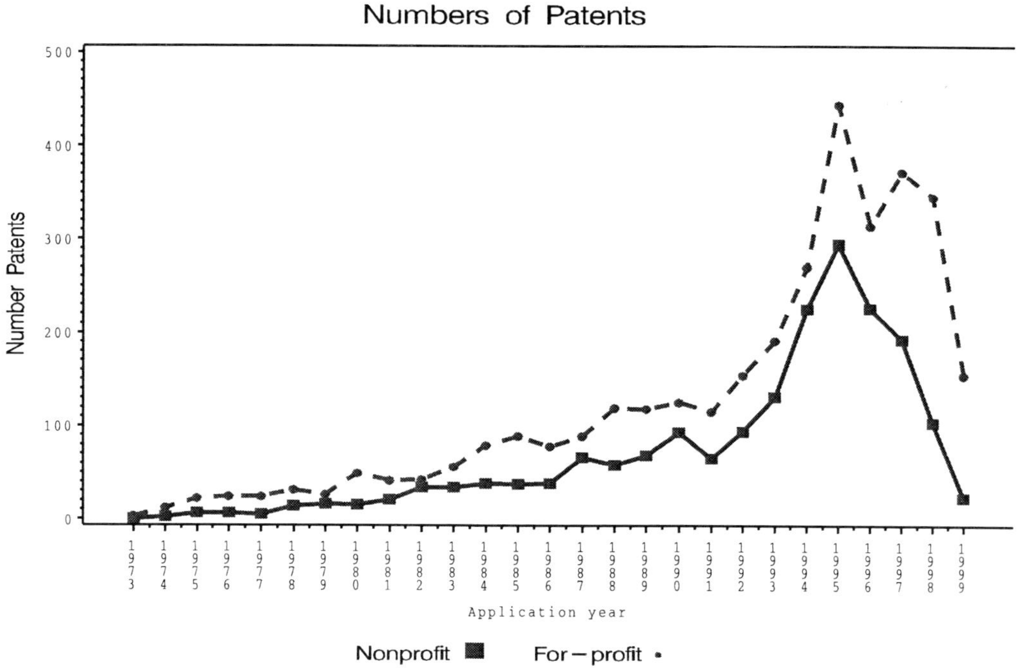

Sources: Author calculations, King and Heisey ABIP (2004).

Figure 1. **Number of Patents by Ownership Type** Total number of patents issued by ownership type and application year. Since the figure only includes patents issued by the year 2000, the decline in the number of issued patents applies for past 1994 suffers from data truncation.

Figure 2 illustrates the average number of successful patents applied for over all firms or organizations by application year and firm type. That is, across all firms in a particular application year, the figure plots the average number of successful patent applications. According to the figure, it appears that the average nonprofit applies for more successful patents than the average for-profit. The difference in the number of successful patent applications between the average nonprofit and the average for-profit firm appears to widen in the 1990s. Repeated measures ANOVA is used to test the null hypothesis that the mean number of (successful) patents owned by nonprofit organizations is the same as the mean number of (successful) patents owned by for-profit firms. Repeated measures ANOVA examines repeated measures of the same dependent variable for subjects and tests for equality of the means between groups. The dependent variable is the number of issued patents applied for by a firm or organization which is observed repeatedly over application years 1976-1997 [12]. There are a total of 688 observations: 200 nonprofit and 488 for-profit organizations. The null hypothesis is rejected. That is, a statistically significant [13] difference exists between the mean number of successful patents owned by a nonprofit organization and the mean number owned by a for-profit firm.

[12]The years 1998 and 1999 are removed to avoid aforementioned data truncation problems.

[13]The null hypothesis is rejected at the 10% significance level.

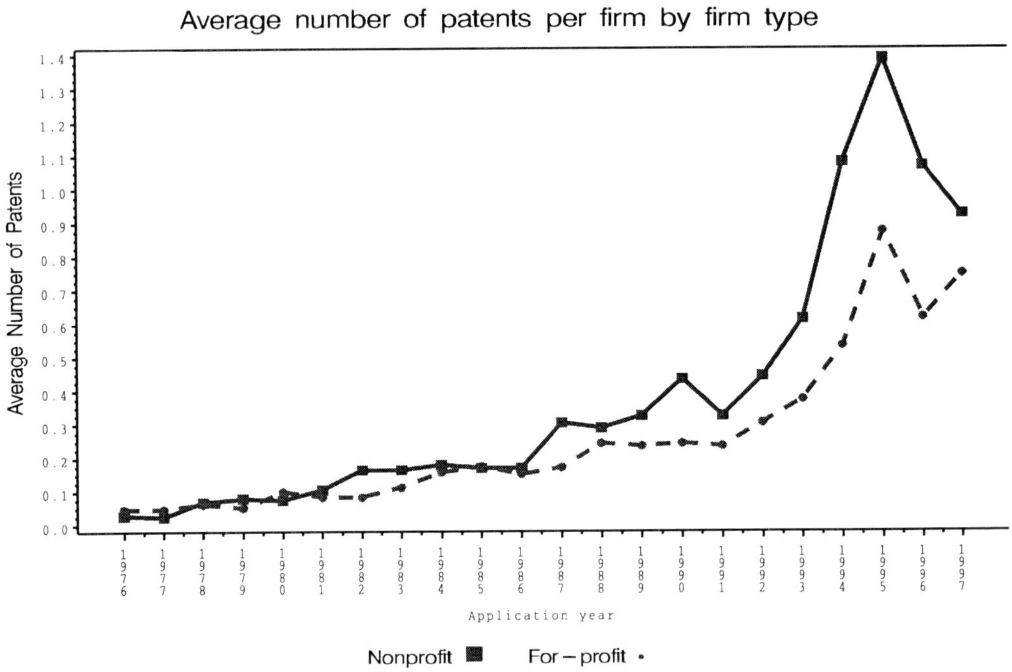

Sources: Author calculations, King and Heisey ABIP (2004).

Figure 2. **Average Number of New Issued Patents by Ownership Type** Average number of new issued patents applied for by a nonprofit organization versus a for-profit firm by application year.

However, these results must be interpreted with caution. First, since no explanatory variables are used, differences may be attributed to a variety of reasons. For instance, nonprofits may have better access to funding. As well, they may have greater numbers of better trained employees (e.g., greater numbers of Ph.D. scientists and access to graduate students). Further, the amount of innovation, captured by a patent count, for a typical for-profit firm may be understated depending on the definition of a "firm." Patent ownership is identified by its assignee name, and assignees may be subsidiaries of larger companies. Depending on one's definition of a firm or entity, the average for-profit parent company may create greater numbers of innovation than that captured by the number of patents associated with its assignee name[14]. Finally, for-profit firms that declare bankruptcy will drag the mean number of patents owned by for-profits down.

Economists have criticized the use of the number of patents as a measure of innovation. A number of alternative measures have been suggested. For example, Hall and Trajtenberg (2001) create a citations-weighted index of patents. However, King and Schimmelpfennig (2005) note that for the ABIP sample of patents, a citations-weighted index leads to a bipolarization of the sample. That is, very few patents are cited very frequently whereas many

[14]The data also does not account for mergers and acquisitions. For example, Monsanto acquired Dekalb in 1996. However, patents continued to be awarded under the Dekalb name after the merger. For more information on recent mergers and acquisitions, see RAFI, 2000.

Table 2. Quantile Statistics for Patents by Ownership Type 1976-2000. Descriptive statistics for all patents issued between 1976-2000.

	Patents owned by U.S. Nonprofits		Patents owned by U.S. For-profits	
Statistic	Number Claims	Number Tech Cat	Number Claims	Number Tech Cat
Max	198	7	190	9
Quantile 3	22	2	22	3
Median	13	1	13	2
Mean	16.6	1.8	17.4	2.1
Quantile 1	7	1	8	1
Min	1	1	0	1
Num Obs		1818		3301

Source: King and Heisey ABIP (2004)

patents are rarely cited. Instead, they suggest using a technology class weighted index. The differences in the patents owned by nonprofits and for-profits are examined by both the numbers of technology classes spanned and the number of claims. Table 2 compares quantiles and statistics for patents owned by nonprofits versus for-profits. According to the table, patents owned by nonprofits and for-profits do not appear to differ much by the number of claims. Also, patents owned by for-profits span only slightly more technology classes on average than patents owned by nonprofits.

Repeated measures analysis of variance (ANOVA) is used to determine whether nonprofits appear to specialize in innovations of a particular type of technology more than for-profits. Here the six broad categories, defined earlier in the text, are used as defined in King and Heisey's ABIP database (2004) to designate the particular types of technology. The number of successful patent applications in a particular technology class is used as a measure of the firm's specialty in a particular technology. An observation is a nonprofit organization or for-profit firm. There are a total of 688 observations: 200 nonprofit and 488 for-profit. The number of new successful patents applied for in a technology class in each application year 1976-1997[15] is calculated for each firm. Repeated measures ANOVA is used to test the null hypothesis that the mean number of successful patent applications owned by nonprofit organizations in a particular technology class is the same as the mean number of successful patent applications owned by for-profit organizations in a particular technology class. The analysis is repeated for each of the six technology classes. For similar aforementioned reasons, the results should be interpreted with caution.

A nonprofit is significantly more likely to own patents than a for-profit in three technology classes: protection, nutrition, and biological control of plants [16]; metabolic pathways and biologic processes, DNA-scale[17]; and genomics[18]. The latter two categories are more related to basic research. For these types of innovations, their commercial values are likely more difficult to assess, and they are farther away from certain commercial development.

[15]The years 1998 and 1999 are removed to avoid aforementioned data truncation problems.

[16]Significant at the 10% significance level.

[17]Significant at the 5% significance level.

[18]Significant at the 5% significance level.

For-profits may be less attracted to creating these technologies since the commercial gains are not immediate. Since commercial development could be far into the future, a patent may expire before the firm can receive any financial benefits from receiving the patent. On the other hand, as previously mentioned, nonprofits and for-profits may also be engaging in patent races, as in the race to map the rice genome. (Naseem and Oehmke, 2004)

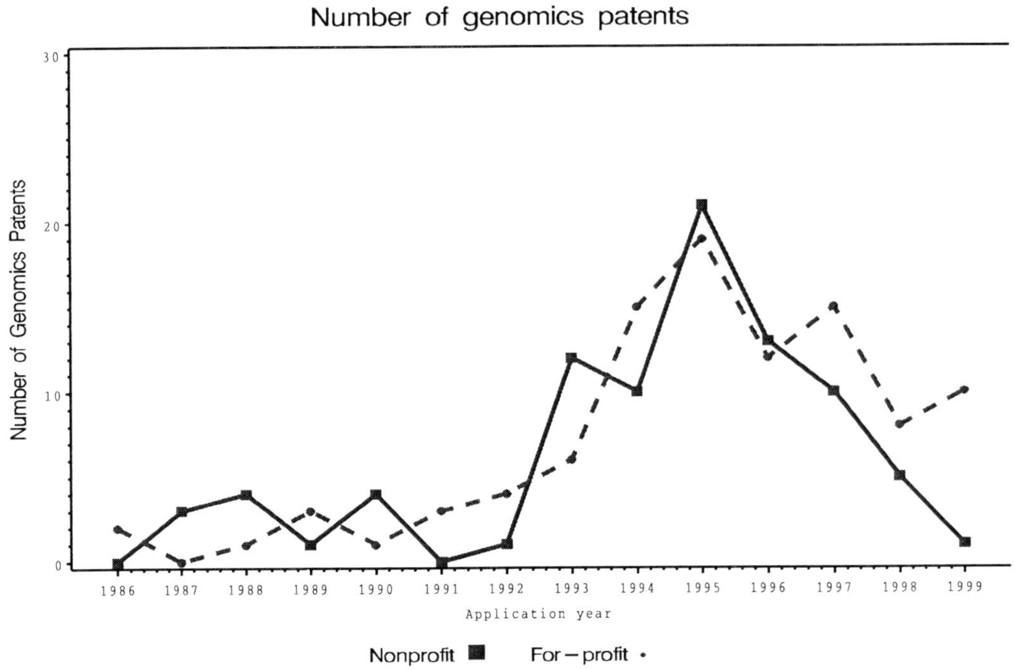

Sources: Author calculations, King and Heisey ABIP (2004).

Figure 3. **Number of Genomics Patents by Application Year** Total number of genomics patents issued by ownership type and application year. Since the figure only includes patents issued by the year 2000, the decline in the number of issued patents applied for past 1994 suffers from data truncation.

Figure 3 illustrates the sum of genomics patents owned by for-profits and nonprofits. Altogether nonprofits and for-profits appear to own roughly equal numbers of genomics patents. For-profit firms are less likely to own any patents in this technology class. Specifically, according to the data, only 55 out of 488 for-profits own genomics patents while 46 out of 200 nonprofits own genomics patents. Relative to the number of patents in other technologies, generally few patents have been awarded in this technology area in the industry. The major owners of genomics patents are the University of California (with 9 total patents awarded by 2000), Pioneer Hibred[19] (with 13 total patents awarded by 2000), and Dupont (with 7 patents awarded by 2000).

[19]Pioneer merged with Dupont in 1999.

5. Plant Variety Protection Certificates

An alternative measure of innovation in the industry is the number of plant variety protection (PVP) certificates. PVP certificates are used to provide a count of the number of new varieties created in the industry[20]. In some sense, an examination of PVP certificates provides an approximate count of the end product for agbiotech. That is, firms seek to characterize genes and discover methods of genetic engineering in order to create new plant varieties. Varieties protected under PVP certificates may have either been created from traditional cross-breeding methods (i.e., cross-pollination) or through genetic engineering. In the sample, there are a total of 79 U.S. nonprofit organizations and 340 U.S. for-profit firms.

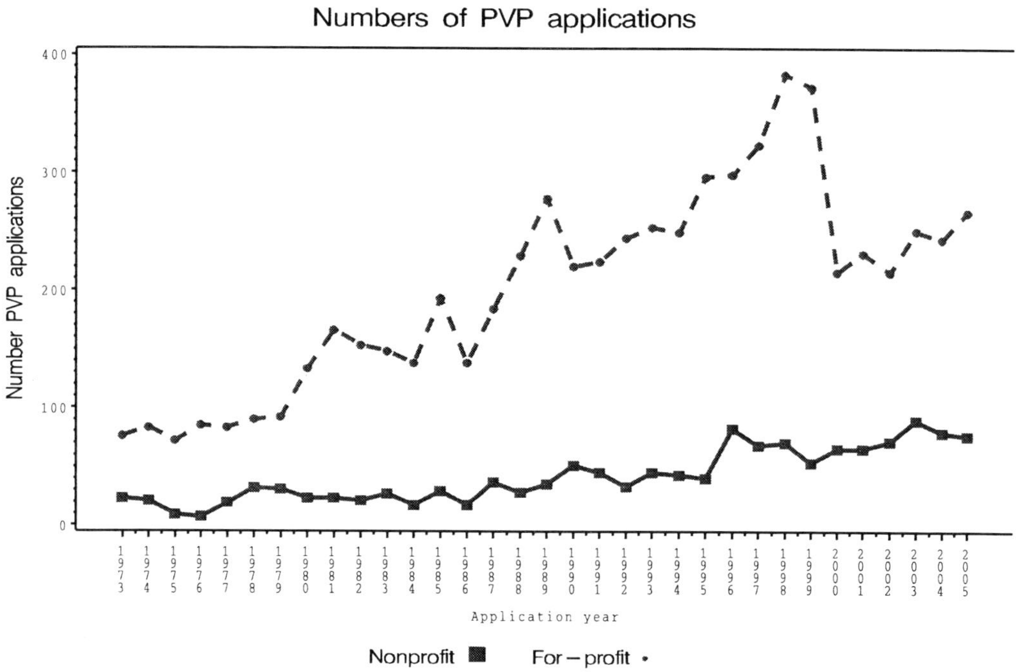

Sources: Author calculations, PVP Office, USDA-AMS (2006).

Figure 4. **Number of PVP applications by Ownership Type** Total number of PVP certificates applications.

Figure 4 illustrates the number of PVP certificate applications by application year. Clearly for-profit organizations have altogether applied for more PVP certificates than nonprofits. However, Butler and Marion (1985) note that since the passing of the PVP Act in 1970, agricultural experiment station directors have felt pressure to release fewer new varieties and instead concentrate more highly on basic research. Nonprofits appear to experience a decline in total PVP certificate applications in 1976, but this is short-lived.

PVP certificates applied for by for-profits increase markedly in several years: 1980,

[20]New varieties may also be protected by utility patents, plant patents, or trade secrets. Therefore PVP certificates provide only an approximation of the number of new sexually reproduced (by seed) or tuber-propagated varieties created.

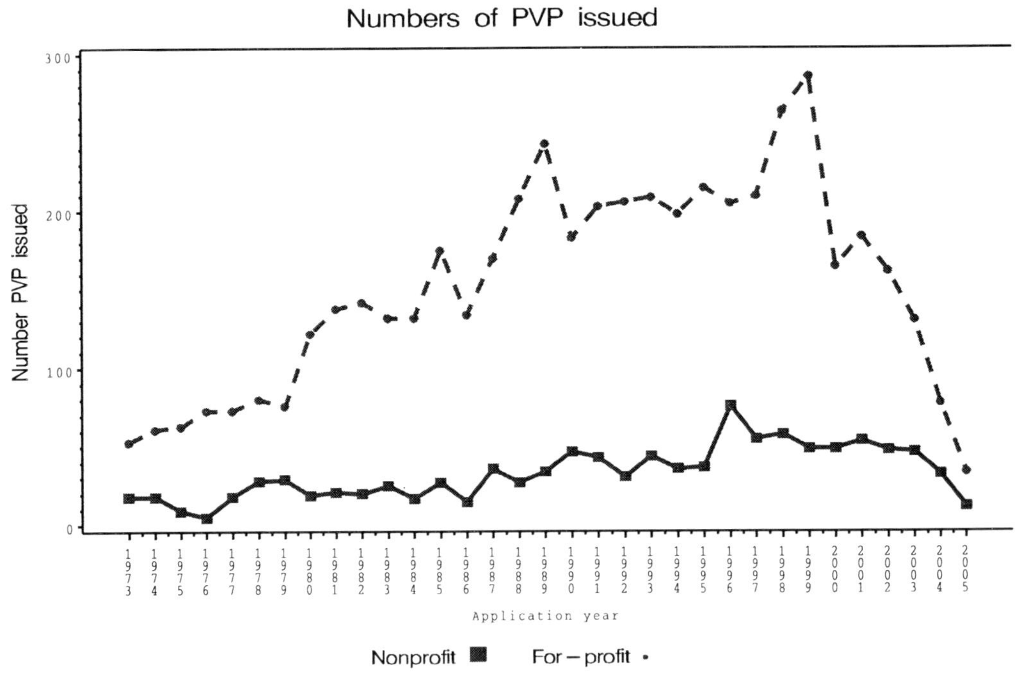

Sources: Author calculations, USDA-AMS (2006).

Figure 5. **Number of successful PVP applications by Ownership Type** Total number of issued PVP certificates by application year. Since the figure only includes certificates issued by the year 2006, the decline in the number of issued certificates in the later years suffers from some data truncation.

1986, and 1995 to 1998. PVP applications from for-profit organizations experience a significant decline between the years 1999 and 2000. Nonprofits do not experience the same decline. The general decline in for-profit PVP applications may be a negative after-effect of the mergers and acquisitions that have occurred among firms in the past ten years. On the other hand, this decline may be an indication that firms are choosing to protect their new varieties with patents rather than PVP certificates (see for example Janis and Kesan, 2001). Since King and Heisey's ABIP database does not extend past patents issued in 2000, this is difficult to verify. However, according to the ABIP database, the number of successful variety patent applications increases steadily throughout the years. See Figure 6.

Figure 5 illustrates the number of *issued* PVP certificates by application year. Again, due to data truncation, the number of issued PVP certificates decline in the later years (2003-2005) since more PVP applications are more likely to be pending approval. Comparing Figures 4 and 5, there do not appear to be great differences between for-profits and nonprofits in the percent of PVP applications that are eventually approved. However, for certificates applied for during the 1990s, for-profits had slightly greater success in receiving approvals than nonprofits.

For-profit participation in the creation of new plant varieties should depend on the perceived profitability of particular varieties. Butler and Marion (1985) list market value of

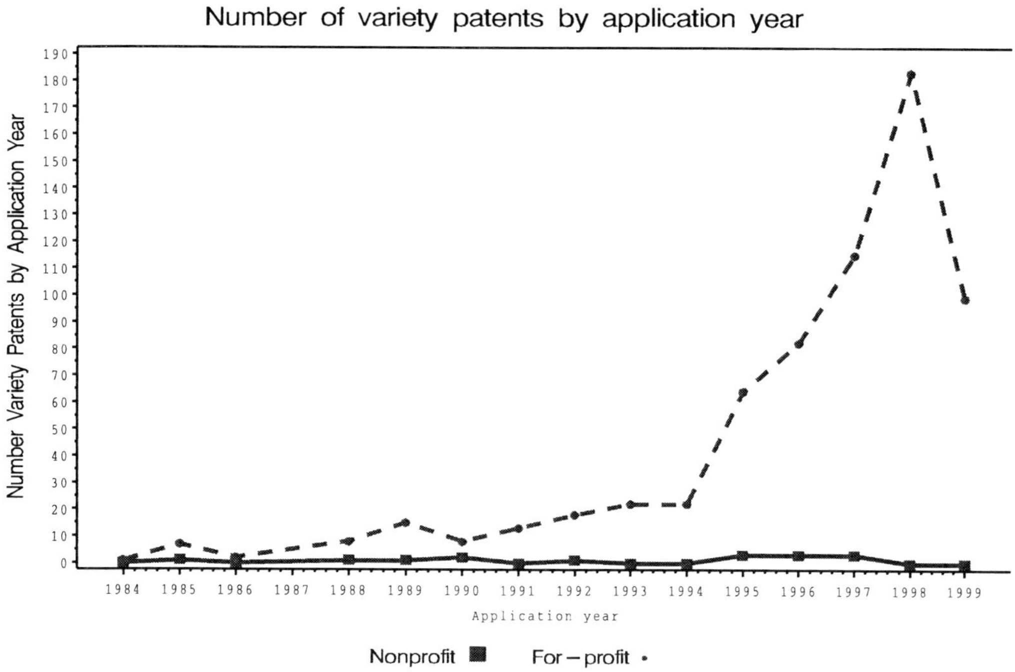

Sources: Author calculations, King and Heisey ABIP (2004).

Figure 6. **Number of Variety Patents by Application Year** Total number of variety patents issued by ownership type and application year. Since the figure only includes patents issued by the year 2000, the decline in the number of issued patents applied for in the later years suffers from data truncation.

production, total quantity of seeds sold, possibility for genetic improvement, ability to mark-up prices, and ease and costs of developing new varieties as determinants of the commercial value of creating new plant varieties. Alston (2004) also discusses several other determinants of profitability as related specifically to agbiotech. These include benefits to farmers, market acceptance, and environmental benefits and costs. For illustrative purposes, Table 3 lists one of the determinants affecting commercial profits: the value of production for selected U.S. crops. Differences in production value could provide one explanation for the observed differences in the creation of particular types of varieties between nonprofits and for-profits. According to Table 3, corn and soybean have among the highest production values. Wheat has lower production value than corn and soybean but higher production value than cotton. However, wheat varieties tend to be very region-specific which could make the creation of a new wheat variety less profitable to for-profits.

The PVP certificates data is used to examine differences in the types of plant varieties created by nonprofits versus for-profits. For example, for-profits are likely to create new varieties in field crops with high production values such as corn and soybean. Meanwhile nonprofits may choose to concentrate on creating varieties in species where for-profits have not.

Table 4 lists the varieties which receive the greatest number of PVP certificates. It then

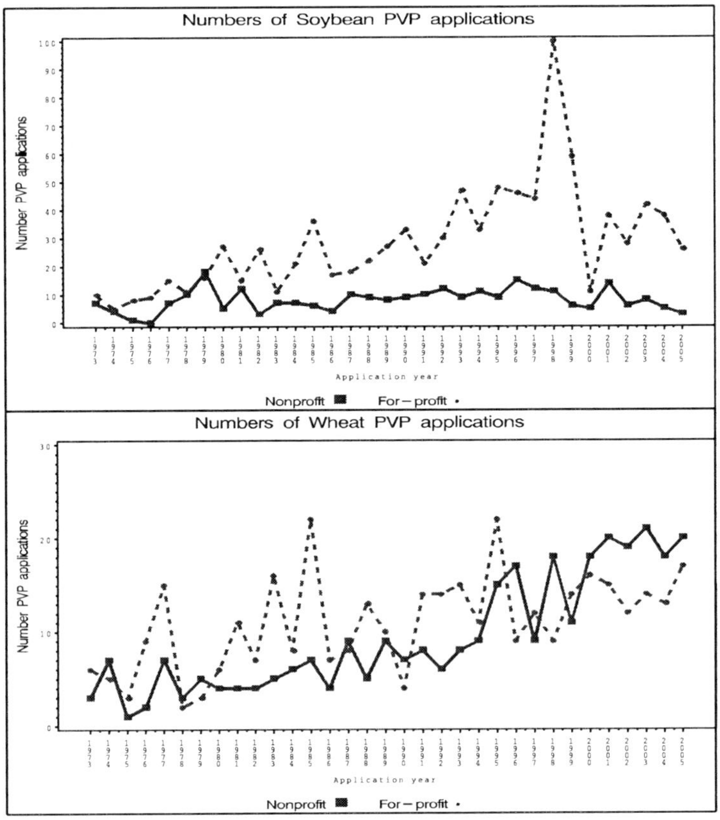

Sources: Author calculations, PVP Office, USDA-AMS (2006).

Figure 7. **Number of PVP applications by Ownership Type for Selected Crops** Illustrates the number of soybean and wheat plant variety protection certificate applications.

divides these into those applied for by nonprofits versus for-profits. Altogether, the industry has created the most new corn, soybean, and wheat varieties. For-profits create over 75% of the new corn and soybean varieties. Meanwhile nonprofits have created nearly half of the new wheat varieties. Finally, only for-profits have played a role in creating new varieties of lettuce. In contrast, plant species for which a significant number of PVP certificate applications originated from nonprofits[21] are oats[22], potato[23], peanut[24], and rice[25]. Rice is a crop considered to be of significant value for developing countries.

The for-profit firms which dominate in the creation of new corn, soybean, and wheat varieties are Pioneer[26] and Monsanto[27]. Delta and Pineland dominates in the creation of

[21]Specifically, where nonprofit organizations have applied for the majority (more than 50%) of all PVP certificates and the total number of PVP certificate applications received exceeds 70 over the years 1971-2005.

[22]There were 71 oat PVP applications: 65 owned by nonprofits and 6 owned by for-profits.

[23]There were 80 potato PVP applications: 52 owned by nonprofits and 28 owned by for-profits.

[24]There were 70 peanut PVP applications: 46 owned by nonprofits and 24 owned by for-profits.

[25]There were 77 rice PVP applications: 41 owned by nonprofits and 36 owned by for-profits.

[26]Acquired by Dupont in 1999.

[27]Some applications are taken out under Monsanto's subsidiaries, Asgrow and Dekalb. For instance, corn

Table 3. Value of U.S. production for selected crops. Lists the the value of production in U.S. dollars for selected crops in years 1999, 2000, and 2001.

	Value of Production 1,000 Dollars		
Crop	1999	2000	2001
Soybean[a]	12,205,352	12,466,572	12,439,597
Corn[b]	17,103,991	18,449,002	19,209,312
Wheat	5,593,989	5,782,107	5,553,815
Cotton	3,809,560	4,260,417	3,383,732
Lettuce	1,431,881	1,871,510	1,907,083
Oats	175,172	175,797	175,923
Peanuts	971,608	896,097	1,003,099
Potatoes[c]	2,745,712	2,591,091	2,933,853
Rice	1,231,207	1,049,961	895,757

Source: USDA, 2002

[a] Soybeans for Beans

[b] Corn for Grain

[c] Includes government payments for loan peanuts, and marketing association payments for pool peanuts in years when there are pool profits.

new cotton varieties, and Seminis[28] dominates in the creation of new lettuce varieties. Of the nonprofits, agricultural experiment stations and public universities own the most PVP certificates for the crops listed in Table 4. For instance, of the nonprofits, North Dakota State University creates the most new corn and wheat varieties while the Mississippi agriculture and forest station creates the most new soybean varieties. However, the nonprofits that create the greatest numbers of new plant varieties create far fewer (less than half the number of) new plant varieties than the most creative for-profit firms (Pioneer, Monsanto, and Delta Pineland).

Figure 7 illustrates the number of PVP certificates applied for by nonprofits and for-profits for soybean and wheat varieties by application year. It provides two different pictures of the changing states of innovation over time for two crop types. While the number of soybean PVP certificates applied for by for-profits generally increases (until 1998), the number of soybean PVP certificates applied for by nonprofits remains fairly steady. In comparison, while the number of new wheat varieties created by for-profits has not declined, the creation of wheat varieties by for-profits has not experienced a similar level of growth as for soybean varieties. Meanwhile the number of wheat varieties created by nonprofits appears to increase over time. Comparing wheat and soybean provides some evidence that nonprofits have increased innovation in a crop type in which for-profits have not expressed a growing interest.

varieties applied for by Dekalb and soybean varieties are applied for by Asgrow.

[28]Owned by Monsanto as of 2005.

Table 4. Top PVP applications crops 1971-2005. Lists the top five crops for which PVP certificates have been applied for by U.S. applicants between the years 1971 and 2005. The last two columns list the number applied for by nonprofits versus for-profits. Percents owned by nonprofits versus for-profits are in parentheses.

Crop	Number	Num nonprofit (%)		Num for-profit(%)	
Soybean	1225	273	(22.3%)	952	(77.7%)
Corn	953	20	(2.1%)	933	(97.9%)
Wheat	688	315	(45.8%)	373	(54.2%)
Cotton	440	73	(16.6%)	367	(83.4%)
Lettuce	347	0	(0%)	347	(100%)

Source: PVP Office, USDA-AMS (2006)

6. Conclusion

This chapter provides some empirical evidence to assess the roles for-profit and nonprofit organizations play in innovation in the U.S. agricultural industry in the last 30 years. Overall, for-profits generally tend to be more "innovative" than nonprofits as measured by the total number of agbiotech patents and the total number of PVP certifications applications and grants held by for-profits versus nonprofits. However, the mean number of patents held by a nonprofit organization is greater than the mean number of patents held by a for-profit firm. Further, nonprofits are more likely to engage in some basic research involving DNA research; that is, a greater share of nonprofit organizations conduct basic DNA research than the share of for-profits. According to the data on plant variety protection (PVP) certificate applications, nonprofits appear to concentrate in certain plant species such as rice. However, oftentimes the creation of new varieties in a particular plant species cannot be defined primarily as either nonprofit or for-profit. That is, nonprofits do not clearly dominate in some crop markets while for-profits clearly dominate in the others. Instead, the share of varieties created by nonprofits ranges anywhere from 0% in some markets to 90% in others[29].

As previous economists have suggested, nonprofits may be providing innovation in areas which for-profits have virtually ignored. On the other hand, nonprofits also pursue some of the same types of innovations as for-profits do. Finally, some would argue that innovation in some varieties is lacking altogether and stress the importance of easier regulatory approvals and higher profit incentives to motivate firms to invest in the creation of particular types of varieties such as fruits and vegetables. (Redenbaugh and McHughen, 2004)

This chapter examined utility patents and plant variety protection (PVP) certificates as measures of innovation. However, these measures ignore the quality of the invention. Therefore it is not known, for instance, the quality and usability of nonprofit innovation in comparison to for-profit innovation. It is well known that many patents are relatively useless. As well, not all new plant varieties are protected with PVP certificates. This chapter

[29]In some plant species, nonprofits create 100% of the new varieties. However in these cases, the total number of PVP certificate applications is less than 10.

also does not discuss the success among nonprofits and for-profits of the development and marketing of new plant varieties.

This chapter focused on describing some available data for measures of innovation and discusses the differences in the magnitudes and types of innovation that nonprofits and for-profits have pursued. However, it invites several unanswered questions. First, the causes of the differences between nonprofit and for-profit innovation remains unanswered. Second, it does not discuss the partnerships between nonprofits and for-profits. For example, the data does not provide details on the amount of licensing between nonprofits and for-profits and the birth of new for-profit companies from nonprofits. Third, it leaves open the discussion of the benefits and costs of having for-profits innovate versus nonprofits and vice-versa. Butler and Marion (1985) list several important roles nonprofits may play in the industry. To some extent, these reasons provide support for nonprofits' participation in the same types of innovation as for-profits. First, nonprofits may provide a check on the prices for-profits may charge for new plant varieties. That is, if nonprofits create varieties that compete with for-profits' varieties, they can prevent for-profits from marking up their prices too high. Second, they may help decrease the negative effects of industry consolidation. That is, if nonprofits maintain ownership of several key patents, the industry is less likely to experience a decline in innovation due to industry consolidation.

On the other hand, Alston (2004) documents many reasons for why for-profits have shied away from creating and developing new varieties for particular plant species. Therefore it remains unclear whether nonprofits are focusing enough attention on the types of innovation ignored by for-profits. Further, nonprofits' incentives have changed over time. With nonprofits' increasing use of exclusive licensing practices, their patent practices are more likely to emulate for-profit patent strategies. Therefore there may no longer be any noticeable differences in the patent behaviors between nonprofits and for-profits.

References

[1] Alston, J. (2004) "Horticultural biotechnology faces significant economic and market barriers." *California Agriculture* **58**(2), 80-88. Available on the World Wide Web at http://californiaagriculture.ucop.edu.

[2] Barton, J. H. (1998) "The Impact of Contemporary Patent Law on Plant Biotechnology Research." *Intellectual Property Rights III Global Genetic Resources: Access and Property Rights*.

[3] Butler, L. and Marion, B. (1985) "The Impacts of Patent Protection on the U.S. Seed Industry and Public Plant Breeding." *Food Systems Research Group Monograph* **16**, University of Wisconsin, Madison.

[4] Graff, G. (1997) "The Agricultural Biotechnology Industry in Overview." working paper. Available on the World Wide Web at http://are.berkeley.edu/ ggraff/ agricultural-biotechnology-industry-in-overview.pdf.

[5] Graff, G., Rausser, G., and Small, A. (2003) "Agricultural biotechnology's complementary intellectual assets." *Review of Economics and Statistics*, **85**(2), 349-63.

[6] Hall, B.H., Jaffe, A.B. and Tratjenberg, M. (2001). *The NBER patent citation data file: Lessons, insights and methodological tools . working paper* **8498**. Cambridge, MA: National Bureau of Economic Research. Available on the World Wide Web: http://papers.nber.org/papers/w8498.pdf.

[7] Janis, M.D., and Kesan, J.P. (2002). "U.S. plant variety protection: Sound and fury...?" *Houston Law Review*. **39**(3), 727-778.

[8] King, J. and Heisey, P. (2004). [database] Available on the World Wide Web at http://www.ers.usda.gov/Data/AgBiotechIP/.

[9] King, J. and Schimmelpfennig, D. (2005) "Mergers, Acquisitions, and Stocks of Agricultural Biotechnology Intellectual Property." *AgBioForum*. Available on the World Wide Web at http://www.agbioforum.com.

[10] Plant Variety Protection Office, Certificate Status Database. (2005). [database] United States Department of Agriculture, Agricutural Market Service. Available on the World Wide Web at http://www.ams.usda.gov/science/PVPO/CertificatesDB.htm.

[11] Mowery, D. (2004) *Ivory tower and industrial innovation: University-industry technology transfer before and after the Bayh-Dole Act in the United States* . Stanford University Press: Stanford, CA.

[12] Naseem, A., Oehmke, J.F., and Schimmelpfennig, D.E. (2005). "Does plant variety intellectual property protection improve farm productivity? Evidence from cotton varieties." *AgBioForum*. **8**(2&3), 100-107. Available on the World Wide Web: http://www.agbioforum.org.

[13] Naseem, A. and Oehmke, J. (2004) "Should the Public Sector Conduct Genomics R&D?" *The Regulation of Agricultural Biotechnology*. R.E. Evenson and V. Santaniello (eds.) CAB International.

[14] Parker, D., Castillo, F., and Zilberman, D. (2001) "Public/Private Sector Linkages in Research and Development: The Case of U.S. Agriculture." *American Journal of Agricultural Economics*. **83**(3), 736-741.

[15] Rural Advancement Foundation International. (2000) "The Seed Giants–Who Owns Whom?" Available on the World Wide Web at http: www.rafi.org.

[16] Redenbaugh, K. and McHughen, A. (2004) "Regulatory challenges reduce opportunities for horticultural biotechnology." *California Agriculture* **58**(2), 106-115. Available on the World Wide Web at http://californiaagriculture.ucop.edu.

[17] United States Department of Agriculture, National Agricultural Statistics Service. (2002) "Crop Values: 2001 Summary." Available on the World Wide Web at http://www.usda.gov/nass/.

[18] Zilberman, D., Yarkin, C., and Heiman, A. (2001) "Privatization and Innovation in Agricultural Biotechnology." working paper. Available on the World Wide Web at http://are.berkeley.edu/∼zilber/DZ/Privatization.pdf.

In: Economics of Technological Change
Editor: Thomas L. Beane, pp. 107-122

ISBN: 978-1-60021-237-6
© 2007 Nova Science Publishers, Inc.

Chapter 6

PRODUCTION EFFICIENCY AND TECHNOLOGICAL CHANGE: A STUDY OF INDIAN RAILWAYS

Anand Venkatesh
Goa Institute of Management, Ribandar,
Goa-403006, Contact-+919822109866

ABSTRACT

Indian Railways (henceforth referred to as IR) is a Department of the Government of India (henceforth referred to as GOI). The control mechanism between GOI and IR leaves IR in a confused situation as far as specification of its objectives is concerned; it is expected to be financially viable while at the same time fulfilling certain social obligations, which are not supported by GOI. Also during the last few years, financial support from GOI has been on the decline thereby adversely affecting the financial position of IR and leaving it to fend for itself by means of market borrowings. In such an environment, it is imperative that IR operates efficiently, i.e. use the inputs at its disposal to get maximum output from the same. IR initially had nine zones many of which were further split to form sixteen zones. GOI and IR justified this division of zones in terms of increase in efficiency due to smallness of size. This paper attempts to find an empirical justification for bifurcation of the zones. In this paper the Data Envelopment Analysis (DEA) technique has been used to study the productive efficiency of IR. This is essentially a non-parametric technique, which does not require restrictive assumptions of cost minimization, which was essentially a limitation even with flexible form cost functions such as the transcendental logarithmic (translog) cost specification. Specifically, since panel data (for nine zones and time period 1989-90 to 1999-2000) was available, the Malmquist DEA technique was used to decompose Total Factor Productivity (TFP) into two components; one part due to efficiency change, which indicates to what extent firms are getting closer to the frontier and technical change, which captures shifts in the frontier itself. In the context of IR, technological change would mean changes in quality of track, wagon and locomotives with respect to production efficiency. This exercise indicated that zones of IR are not making optimal use of the inputs at their disposal. Also larger zones of IR, in terms of output (passenger kilometres and ton kilometres), seemed to be relatively efficient as compared to smaller zones. This seriously challenges GOI's decision to split IR into smaller zones, which seems to be devoid of any economic rationale. Another interesting fact emerging from the study was that while zones of IR made the best possible use of whatever inputs were made available to them, IR as an organization has done precious little in embracing cost minimizing technology.

INTRODUCTION

The Indian Railways (henceforth referred to as IR) is among the largest transport undertakings in the world under a single management and has conventionally been operating as a departmental undertaking of the Government of India (henceforth referred to as GOI). IR is a central government monopoly with a mandate not only to provide rail transport services, but also to fulfill certain social obligations. Its revenues account for about one percent of India's GDP. The organization employs around 1.6 million persons. The rail network is extensive; with a route length of around 63000 kilometres of which nearly 44000 kilometres are broad gauge sections and the rest are mainly metre gauge sections. IR is managed by a Railway Board and is divided into nine zonal railways.[1]

IR could be classified as what is known in economic literature as a multiproduct enterprise, i.e. producing more than one output. Broadly speaking, IR moves passenger as well as freight traffic, thereby producing both Passenger Kilometres (PKM) as well as Ton Kilometres(TKM). However, even within these segments, it provides differentiated services. The system provides intercity freight as well as suburban commuter rail services in major cities. In addition, IR owns and operates almost all rail transport support services including design and manufacturing of rolling stock, rail catering, schools, technical institutes, housing, etc.

IR is currently finding that it is operating in an increasingly competitive market, which further necessitates the need to pay careful attention to its pricing policy since, over the years, a large proportion of traffic (especially freight) has been diverted from rail to roads. Unlike the roads sector, where numerous qualified motor vehicle operators can operate anywhere on the road network, the rail sector is a restricted transport system whose vehicles are confined to a fixed guide way. In both the passenger as well as freight segment, IR is facing stiff competition from roads for both short as well as long distance traffic. This is despite the fact that road movements face significant diseconomies of scale over long distances. Also, rail transport is considered to be energy efficient as compared to road transport. During recent decades and more so in the 1990s, IR has been forced to function as a commercial undertaking while at the same time required to fulfill certain social obligations. Thus, IR today could be said to be an unenviable position wherein due to dwindling financial support from GOI, it has been forced to find alternative sources of finances (Sriraman, 2002)

The fact that IR is a departmental undertaking of GOI has in itself been a cause of several distortions in production, pricing and investment. The Tenth Plan (GOI, 2002) insists that IR operate along commercial principles. at the same time, the popular view is that IR is a public utility, which is expected to fulfill certain social obligations. This has resulted in IR adopting certain populist pricing policies. It is believed, and with good reason that the pricing policies adopted by IR has led to distortions in resource allocation and distribution both at the organizational as well as at the macro economic levels. Anand (1998) observes that the average cost per PKM increased due to improvement in quality of passenger trains and the average cost of transporting freight came down due to IR's policy of eliminating wagon and part wagon load and hence the associated shunting and marshalling operations. However, this

1 In the year 2001, from the existing nine zones of IR, new zones were created and currently IR has sixteen Zones.

was not evident in IR's pricing policy wherein the ratio of the rate per PKM to the rate per TKM showed a decline rather than appreciation over the years.

Of equal concern for IR is the congestion along certain high density passenger corridors such as the golden quadrilateral[2] due to the relatively low prices charged. This has also led to several inefficiencies in production. Of IR's total route kilometrage, only 15000 km are double or multitrack. Not only is IR facing capacity constraints on its high-density corridors, but its operations have also suffered due to maintenance backlog over the last ten years or so. Arrears in track renewals have increased from 3,548 km in 1991-92 to more than 12,260 km in 2001-02. Inefficient maintenance of rail tracks and rolling stock has led to more accidents, train derailments, higher operating costs and loss of revenue from suspended operations on the affected lines. For the past 20 years, freight trains have run at an average speed of 23 km per hour. With electrification and modern locomotives, the average speed should have increased to 40-50 km an hour. These could act as indirect evidence to indicate that IR is not operating efficiently, i.e. not producing its output at minimum possible cost. This is notwithstanding the fact that there have been several improvements in various productivity indicators during the Ninth Five Year Plan period (GOI, 1999). Given that IR may not be producing efficiently, i.e. producing the maximum output with its given inputs. It is imperative that the levels of production efficiency of IR be computed so that one is in a position to ascertain whether and to what extent there is a need for IR to explicitly incorporate and pursue the objective of cost minimization. This in essence is the objective of the study. Also, recently, IR has created smaller Zones from its existing Zones and cited the reason of enhancing production efficiency due to smallness in size for the same. This study tries to determine to what extent this hypothesis is empirically justified.

Earlier Studies of productivity and efficiency either adopted a partial factor productivity approach or a cost function approach (Jha et al, 1993). However, a serious limitation of the cost function approach is the assumption of efficient production inherent in the construction of the cost function. For undertakings like IR, which seem to have inefficiencies in production due to several reasons, the assumption of efficient production, i.e. operating on the "efficiency frontier" is rather heroic. Hence one needs to seriously consider alternatives to the cost function approach.

An approach, which estimates production efficiency without requiring the stringent assumptions of cost minimization, is the non-parametric Data Envelopment Analysis (DEA) approach. This is the technique that has been used to estimate the levels of inefficiency prevailing in IR.

This study proceeds as follows; the next section focuses on certain conceptual issues regarding production efficiency as well as a description of the Malmquist DEA approach. This is followed by description of the data relating to inputs and outputs in the context of IR in Section 3. Section 4 presents the analysis of various results obtained using the Malmquist DEA approach. Section 5 provides the summary and policy recommendations based on the earlier analysis. Finally Section 6 points out some possible limitations of the study and areas for further research.

---

2 The "Golden Quadrilateral" is the corridor connecting the four prominent Indian cities of Delhi, Mumbai, Kolkota and Chennai.

2. PRODUCTIVITY AND EFFICIENCY: CERTAIN PRELIMINARY CONCEPTS

As seen above, the main objective of the study is to determine the levels of inefficiency prevailing in IR. It would be useful to understand efficiency, productivity and other related concepts in order to get a better insight into the same. It is also imperative to understand the technique used to measure efficiency viz. the Malmquist variant of the Data Envelopment Analysis (DEA) technique. The literature review is by no means comprehensive but the main attempt here is to introduce all the main concepts so as to enable the uninitiated reader to get a clear picture of the same before proceeding to elaborate the methodology and data used to estimate relative efficiency levels in IR.

The productivity of a firm is defined as ratio of the output(s) that it produces to the input(s) it uses. Consider the case of a firm producing only a single output (y) using a single output (x). Productivity of this firm is defined as

$$\text{Productivity} = y/x \tag{1}$$

As can be seen measurement of productivity is quite obvious in this case. However, in the case of the firm producing multiple outputs using multiple inputs, computation of productivity becomes slightly more complicated. This is particularly true for multiproduct enterprises like IR, which produces two outputs viz. Passenger Kilometers and Ton Kilometers and uses several inputs. It is then necessary to use index number approach to aggregate these inputs and/or outputs into a single input and output index. This is briefly the Total Factor Productivity (TFP) approach, which will be discussed later in this section.

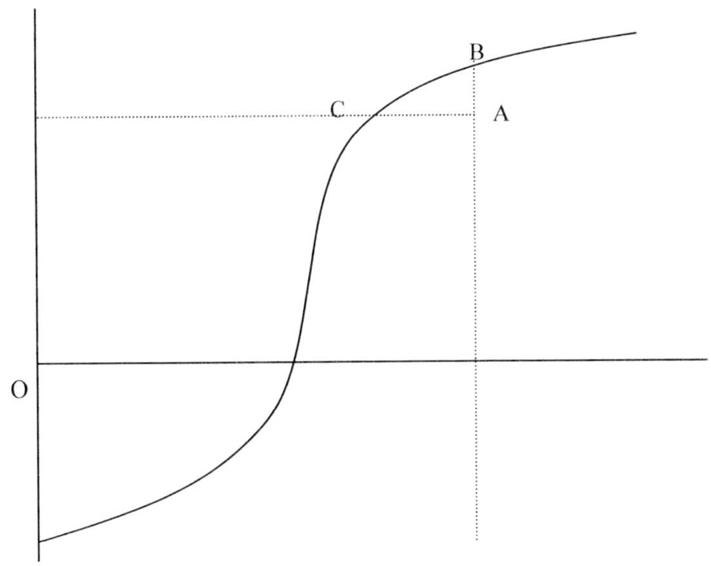

Figure 1. Production Frontier

A term, which is quite often mistakenly used synonymously with productivity, is *efficiency*. In simple terms, efficiency relates to how much output(s) an enterprise is *currently producing* with a given set of inputs with what it *could have produced* adopting best practice production techniques. Figure 1 below shows a graphical exposition of production efficiency. Once again, consider a firm producing a single output using a single input. The input-output combinations depicting best-practice production techniques constitute what is known as the *"production frontier"*[3]

Any point on the production frontier indicates use of best practice production techniques and hence is said to be *technically efficient*. Conversely, any point below the frontier indicates sub-optimal utilization of input and hence is said to be *technically inefficient*. Productivity could be interpreted as the slope of a ray from the origin to a point on the production frontier. Thus as we move from point A to point B in Figure 1, the slope of the curve, and hence productivity increases. Thus, point C, which is a tangent to the production frontier, denotes maximum possible productivity.

One can also consider change in efficiency with time. With the advent of time technology could improve (decline) in the sense that the maximum possible output(s) produced by a given set of inputs could actually increase(decrease). This implies a shift in the production frontier as shown in figure 2. In such a situation technical progress (decline) is said to have occurred. Thus with time two things could happen: one, the firm under study could move closer to the efficiency frontier and two, the efficiency frontier itself could shift with time. This is the intuitive basis to the Malmquist approach for measuring productivity

As mentioned before, if there are more than one input and/or output then measuring productivity and hence drawing inferences upon relative firm efficiencies becomes slightly more involved. Initially, one may be tempted to adopt relatively simple and straightforward techniques for measuring productivity such as *partial factor productivity measures*. These measures focus on the performance of a single factor of production assuming that other factors remain unchanged. In the context of railways, some measures of partial factor productivity could be PKM per rupee spent, TKM per unit of labour, wagon turnaround time, quality of track, etc. Admittedly, these measures are useful in their own right; however, drawing inferences on productivity of factors based on the partial factor productivity scores could lead to serious pitfalls. Consider an example of a bus company, which has experienced steady improvements in fleet utilization and fuel efficiency, thereby implying technical progress. One would be inclined to infer that the improvements in FU is due to improvement in the quality of buses, more fuel efficient engines, etc whereas in reality, none of these might have happened. What could have indeed happened was the bus crew and maintenance staff may have been trained to adopt better driving and maintenance practices so as to enhance the above mentioned factors. More formally, one could mistakenly attribute the technical progress to capital whereas in reality it has been *embodied* in labour. This example illustrates the fact that one must have a good understanding about the undertaking in question while drawing inferences from partial factor productivity measures.

In the more realistic case of multi-input and/or multi-output production, the relevant concept used is Total Factor Productivity (TFP). This is an extension of the one-input, one-output situation. Essentially, TFP is measured as the ratio of output index to input index i.e.

$$\text{TFP} = \text{output index}/\text{input index} \quad (1)$$

---

[3] This section has been based on Coelli et al (1996)

TFP change between two periods, indexed as 0 and 1 could be expressed as:

$$\Delta TFP = \frac{TFP_1}{TFP_0} \quad (2)$$

TFP change can be measured either by using index number approach or the production frontier approach. Each of these approaches has their own merits and drawbacks. The index number approach does not require extensive data. Actually, all that is required for this approach is quantity and price data for just two periods. However, this approach is unable to decompose TFP measure into various components

The frontier approaches fall under two broad categories: The Stochastic Frontier Approach (SFA) and the Data Envelopment Approach (DEA). The SFA is an econometric technique wherein the production function is estimated by splitting the error term into two components; one component capturing the unpredictable elements, i.e. noise, and the other capturing the technical inefficiency. Ideally, panel data covering a number of years is required for the SFA approach. It allows for subjecting the model to statistical hypotheses tests. However, it is subject to the distributional form specified for the error term. Also it requires a relatively large sample size for robust estimation.

Another approach to estimating changes in TFP is using non-parametric methods to build the stochastic frontier. This is what is known as the Data Envelopment Analysis (DEA) technique. This technique essentially is a linear programming method that constructs a nonparametric production frontier by fitting a piecewise linear surface over the data points. We will focus on the DEA technique a little more as it is a variant of this technique that we are going to use to measure changes in TFP. Before looking at the DEA approach let us acquaint ourselves with certain concepts and tools used to measure changes in TFP. However, before doing that, let us familiarize ourselves with a very important tool required while measuring changes in TFP, which is input distance function.

The input distance function defines the distance between a given output vector and an input vector, as the maximum value of a scalar,ρ, such that the scaled input vector, x/ρ remains feasible. i.e.

$$d_i^t(x,y) = \max\{\rho: (x/\rho \in L^t(y)\} \quad (3)$$

Where $L^t(y)$ is the input requirement set i.e.

$$L\ (y) = \{x: x \text{ can produce } y\} \quad (4)$$

Having defined input distance functions, it is now possible to illustrate how the same is applied to measure changes in TFP. We use the input oriented Malmquist productivity index.

The input oriented productivity measure compares the input requirements for producing output level, y_t produced under period t technology, with the input that would have been required if the production technology was the same as that in period s. thus this index compares x_t with what would have been required under technology of period s. Thus period s based input-oriented Malmquist productivity index is defined as:

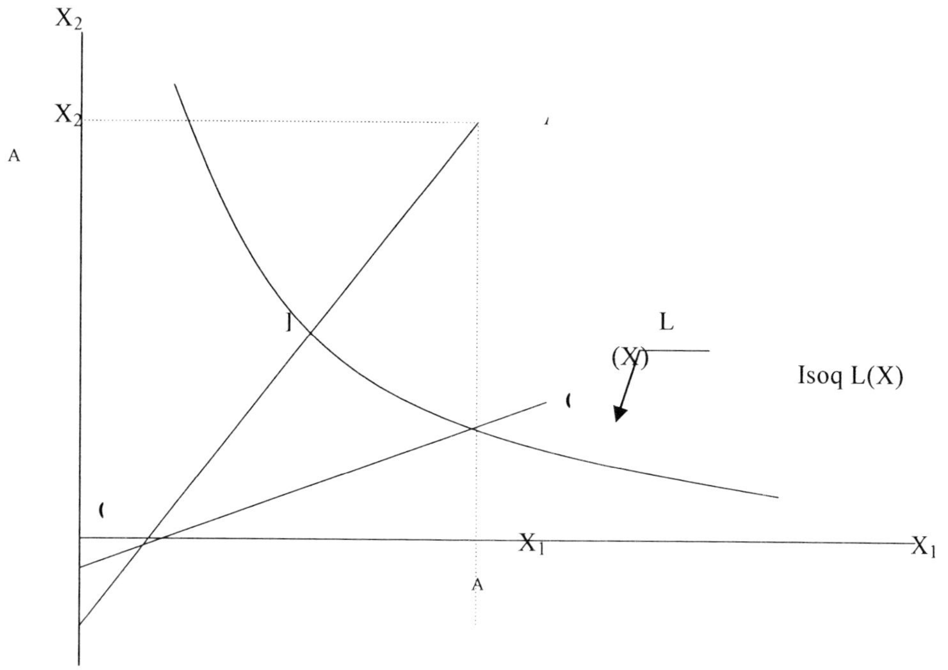

Figure 2. Input Distance Function

$$m_i^s(y_s, y_t, x_s, x_t) = \frac{d_i^s(y_t, x_t)}{d_i^s(y_s, x_s)} \qquad (5)$$

Similarly, we can define period t based input-oriented Malmquist productivity index as :

$$m_i^t(y_s, y_t, x_s, x_t) = \frac{d_i^t(y_t, x_t)}{d_i^t(y_s, x_s)} \qquad (6)$$

Following Fare et. al. (1994), the Malmquist (input oriented) TFP change between period s (the base period) and period t (the current period) is given by:

$$m_i(y_s, x_s, y_t, x_t) = \left[\frac{d_i^s(y_t, x_t)}{d_i^s(y_s, x_s)} x \frac{d_i^t(y_t, x_t)}{d_i^s(y_s, x_s)} \right]^{1/2} \qquad (7)$$

where:

y_s and y_t represent vector of outputs in period s and t respectively
x_s and x_t represent vector of inputs in period s and t respectively

$d_i^s(y_t, x_t)$ represents the distance from the period t observation to the period s technology. A value of m_i greater than one will indicate positive TFP growth from period s to period t while a value less than one indicates a TFP decline.

An equivalent way of writing the TFP index given in eq.7 is:

$$m_i(y_s,x_s,y_t,x_t) = \frac{d_i^t(y_t,x_t)}{d_i^s(y_s,x_s)}\left[\frac{d_i^s(y_t,x_t)}{d_i^t(y_t,x_t)} x \frac{d_i^s(y_s,x_s)}{d_i^t(y_s,x_s)}\right]^{1/2} \quad (8)$$

Where the ratio outside the square bracket measures the change in the input oriented measure of technical efficiency between period's s and t i.e. the efficiency change is equivalent to the ratio of technical efficiency in period t to the technical efficiency in period s. The remaining part of the eq.8 is a measure of technical change. It is the geometric mean of the shift in technology between two periods, evaluated at y_s and also at y_t[4]. Thus the terms in eq.8 are:

$$\text{Efficiency change} = \frac{d_i^t(y_t,x_t)}{d_i^s(y_s,x_s)} \quad (9)$$

And

$$\text{Technical Change} = \left[\frac{d_i^s(y_t,x_t)}{d_i^t(y_t,x_t)} x \frac{d_i^s(y_s,x_s)}{d_i^t(y_s,x_s)}\right]^{\frac{1}{2}}. \quad (10)$$

The DEA estimation using Malmquist technique involves estimation of the four distance functions in eq.8 by mathematical programming and subsequently, computation of TFP change using either eq.7 or eq.8.

The term outside the bracket denotes the "*catch up*" effect, i.e it is a measure of how much closer to the boundary the company is in period t compared to period s. if the catch up term is 1 the company has the same efficient boundary in period t and s from the respective efficient boundaries. if the catch up term is greater than one, then the firm has moved closer to the period t frontier and vice versa for the catch up term being under one.(Thanasoulis,2001)

The term within the bracket represent the *frontier shift* and measures the movement of the frontier between periods s and t at two locations: a frontier shift score of over one represents technical gains by the whole industry in period t over period s and vice versa. It must be understood that the frontier shift to the right represents efficient production in period t as compared to period s as exhibited by the entire industry and not necessarily by any individual firm of that industry whereas a leftward frontier shift represents decline in efficiency for the industry. When the frontier shift term is 1 the industry has on average registered neither a productivity gain nor a productivity loss between periods s and t.

We now examine the methodology of Constant Returns to Scale (CRS) input oriented DEA model developed by Charnes, Cooper and Rhodes (1978). Assuming availability of data

---

[4] The two periods'(s and t) TFP indices are equivalent only if technology is Hicks input neutral i.e. $d_i^t(x,y) = A(t)d_i(x,y)$. we take the geometric mean to avoid the necessity to either impose this restriction or arbitrarily choose one of the two indices as done by Caves, Christensen and Diewert (1982)

on K inputs and M outputs for each of N firms, for the ith firm, these are represented by the vectors x_i and y_j respectively. The (KxN) input matrix, referred to as X and the (MxN) output matrix, referred to as Y represent the data for all N firms. For each firm we would like to obtain a measure of the ratio of all outputs over all inputs i.e. $u'y_i/v'x_i$ where u is an (Mx1) vector of output weights and v is a (Kx1) vector of input weights. The optimal weights are obtained by solving the mathematical programming problem:

$$\begin{aligned} &\text{Max}_{u,v}\; u'yi/v'x_i \\ &\text{Subject to } v'x_{i\,=}1 \\ &U'yj/v'xj \leq 1 \;\; j=1, 2, \text{------}, N \\ &u, v \geq 0 \end{aligned} \tag{11}$$

Equation 3.35 gives the ratio form or the multiplier form of DEA linear programming problem.

Using the principle of duality in linear programming, one can derive an equivalent envelopment form of this problem:

$$\begin{aligned} &\text{Min}_{\theta,\lambda}\theta \\ &\text{s.t. } -y_i + Y\lambda \geq 0 \\ &\theta x_i - X\lambda \geq 0 \\ &\lambda \geq 0 \end{aligned} \tag{12}$$

Where θ is a scalar and λ is a (Nx1) vector of constants. The value of θ obtained will be the efficiency score for the ith firm. It will satisfy $\theta \leq 1$ with a value of 1 indicating a point on the frontier and hence a technically efficient firm. The linear programming problem has to be solved N times, once for each firm and a value of θ is then obtained for each firm.

Intuitively, DEA is explained by figure 3 below.

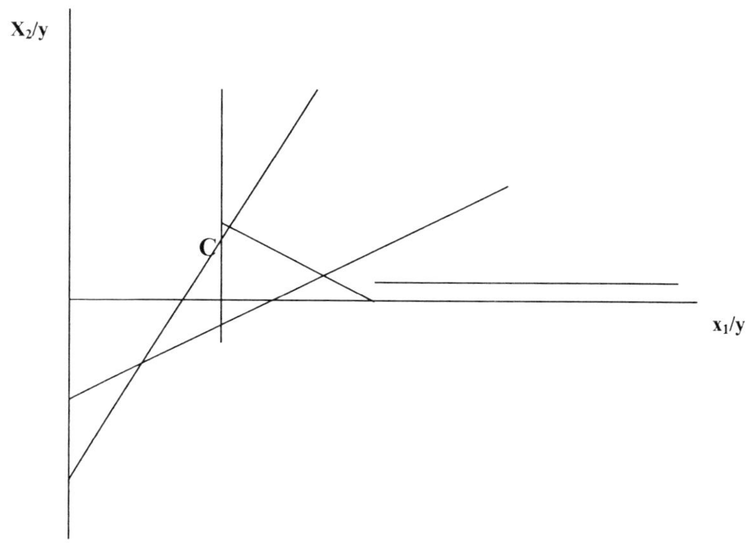

Figure 3. Measurement of technical efficiency

The problem takes the ith firm and seeks to radially contract the input vector as much as possible, while still remaining within the feasible input set. The inner boundary of this set is a piecewise linear isoquant determined by the observed data points. The radial contraction of the input vector, x_i produces a projected point, ($X\lambda$, $Y\lambda$) on the surface of this technology. This projected point is a linear combination of these observed data points. The constraints in equation 3.36 above ensure that this projected point cannot lie outside the feasible set.

Relative technical efficiencies for all zones of IR were estimated using the Data Envelopment Analysis package DEAP (Version 2.1) (Coelli, 1996)

Assuming that panel data is available, we follow Fare et al(1994) and formulate DEA like linear programs with input orientation[5]. This involves calculating four distance functions to measure the TFP change between the two periods. Thus four linear programming problems need to be solved. Assuming a CRS technology, the required LPs are[6]:

$$d_i^t(y_t, x_t) = \min_{\theta,\lambda} \theta$$

Subject to $-y_{it}+Y_t\lambda\geq0$ (13)
$\theta x_{it}-X_t\lambda\geq0$
$\lambda\geq0$

$$d_i^s(y_t, x_t) = \min_{\theta,\lambda} \theta$$

Subject to $-y_{is}+Y_s\lambda\geq0$ (14)
$\theta x_{it}-X_s\lambda\geq0$
$\lambda\geq0$

$$d_i^t(y_s, x_s) = \min_{\theta,\lambda} \theta$$

Subject to $-y_{is}+Y_t\lambda\geq0$ (15)
$\theta x_{it}-X_t\lambda\geq0$
$\lambda\geq0$

and

$$d_i^s(y_t, x_t) = \min_{\theta,\lambda} \theta$$

Subject to $-y_{is}+Y_s\lambda\geq0$ (16)
$\theta x_{it}-X_s\lambda\geq0$
$\lambda\geq0$

since production points are compared to technologies from different time periods, the θ parameter need not be less than or equal to one. The data point could lie below the feasible production set. This will most likely occur in eq.16 where production in period t is compared

5 Input oriented DEA is chosen organizations like IR are obliged to produce a specified output level, hence the problem is one of doing so with minimum possible inputs.

6 The CRS technology is generally assumed to be preferable to the VRS technology while computing Malmquist efficiency scores (Fare et. al., 1994)

to production in an earlier period s. θ could be greater than one if technical progress has occurred between both periods.[7]

Although changes in TFP could be computed using the VRS assumption, it is advisable, considering computational and related problem, to compute TFP changes under constant returns to scale assumption.

3. DATA AND RELATED ISSUES

Panel data was used for estimating the cost function. Data was collected for all nine zones of IR for the period 1989-90 to 1999-2000. This period is relevant in the Indian context as the liberalisation of the economy started in 1991, though the direct impact of the same on IR is at best, uncertain. Two output measures were considered viz. passenger and freight output measured in terms of billion Passenger Kilometres (PKM) and billion Ton Kilometres (TKM) respectively. The inputs used were labour fuel and capital.

The total number of employees was used for labour. IR uses a variety of fuels such as diesel, electricity and coal. Over the last decade or so, the fuels predominantly used were diesel and electricity. Diesel quantity was expressed in kilolitres while electricity usage was given in kilowatt-hours. Thus, we had to formulate a quantity as well as a price index. Both kilowatt-hours and kilolitres were converted into ton oil equivalent (TOE)[8] and thus the index for fuel consumption by zones of IR was formulated.

The total investment during a particular year was taken as proxy for capital since it was otherwise difficult to compute a physical index for capital. The Investment values were adjusted for inflation and hence the real values were reported

The main source of data was Annual Statistical Statements (ANS) of the Indian Railways.

The package used by us for computing TFP change using Malmquist DEA approach was the DEAP software, version 2.1 by Coelli(1999).

4. KEY FINDINGS

Application of the Malmquist DEA technique to IR produced several interesting results. As shown in Figure 4, Taking 1989-90 as the base period of comparison and hence assigning a TFP score of unity to the same, it can be seen that with reference to the base year, TFP grew quite sharply during the year 1990-91 and thereafter remained virtually remained stagnant throughout. Even after a decade the overall TFP growth was only to the extent of 5.8%. The catch up effect was only to the extent of 2.6% over the sample period while the frontier shift was 3.1%. This indicates that IR, as an organization has not pursued an active efficiency maximizing or cost minimizing policy.

---

7 such a scenario could also occur in eq.15 if technical decline occurred but is less likely.

8 One kilolitre of diesel = 0.8639862 TOE (Source: Indian Railways; Annual Statistical Statements)
One Kilowatt hour of electricity=$220*10^{-6}$ TOE (source: http://ncb.intnet.mu/cso/medrc/energy/intro.htm in third issue of the Economic and Social Indicators on Energy Statistics, Republic of Mauritius

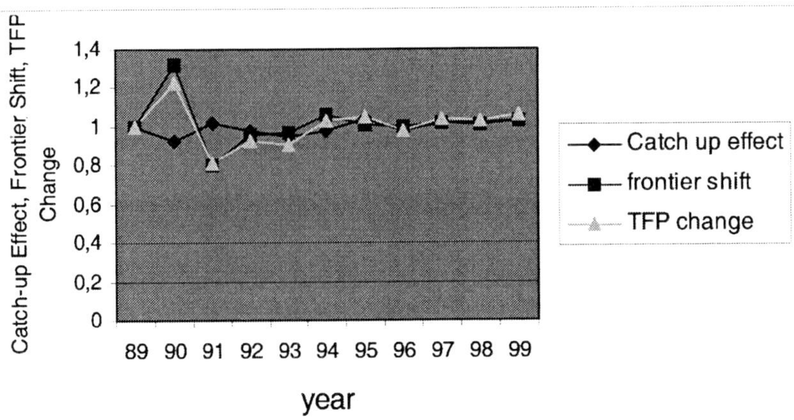

Figure 4. Productivity change for IR from 1989-90 to 1999-2000 using Malmquist approach

The average changes in catch up effect, frontier shift as well as TFP are as given in Table 1. As can be seen, the average change in TFP for each of the nine zones of IR is indeed quite small. While the individual zones, apart from Central, Southeastern and Western Zones, seem to drift away from the efficiency frontier, as shown by the drop in the catch up effects of the respective zones. However, this fall has been quite low, thereby indicative of the fact that the zonal managers of IR have used the inputs available at their disposal to stay as close to the frontier as possible. However, the lack of a pronounced shift in the production frontier provides further evidence of the fact that cost minimization has not been the top priority in IR's agenda.

Table 1. Average of Productivity Indices of IR (Period: 1989-90 to 1999-2000)

Zone	Catch up effect	frontier shift	TFP change
Central	1	0.998	0.998
Eastern	0.989	0.954	0.944
Northern	0.986	1.038	1.023
N.Eastern	0.985	1.01	0.995
Northeast Frontier	0.971	0.982	0.953
Southern	1.002	1.038	1.04
South Central	0.999	1.042	1.041
South Eastern	1	0.99	0.99
Western	0.992	1.042	1.033

Table 2 presents the various statistics related to Malmquist index both on an annual basis as well as for the entire sample period viz. 1989-90 to 1999-2000. Only two zones exceeded the third quartile Malmquist score for the entire sample period. It can also be seen that there appears to be years in which the Malmquist indices have shown impressive growth but the same has not had a sustained effect over the sample period. Over the sample period only four zones show increase in TFP, the maximum being 4.4% for Southern Zone, which by any standards is moderate.

Table 2. Distribution of Malmquist Productivity Indices[9]

year	Q1	Q2	Q3	M_i>1
1990-91	0.99	1.018	1.034	4
1991-92	0.952	1.027	1.038	6
1992-93	0.913	0.936	0.945	1
1993-94	0.903	0.96	0.975	1
1994-95	0.997	1.031	1.065	5
1995-96	1.022	1.066	1.099	7
1996-97	0.948	0.991	1.011	3
1997-98	1.009	1.052	1.095	7
1998-99	0.989	1.047	1.082	6
1999-2000	1.037	1.055	1.08	10
1989-2000	**0.99**	**0.998**	**1.033**	**4**

Q1- First Quartile
Q2-Median
Q3-Third Quartile

In order to understand the Malmquist TFP change better, one would like to analyze it with respect to the DEA technical efficiency scores of the base year viz. 1989-90. this is shown in figure 5. it is observed that there is a moderate positive correlation of 0.47, which is however not statistically significant. Thus it cannot be said with certainty that there is a relationship between how well a firm was doing in terms of technical efficiency in the base year and the TFP change over the sample period.

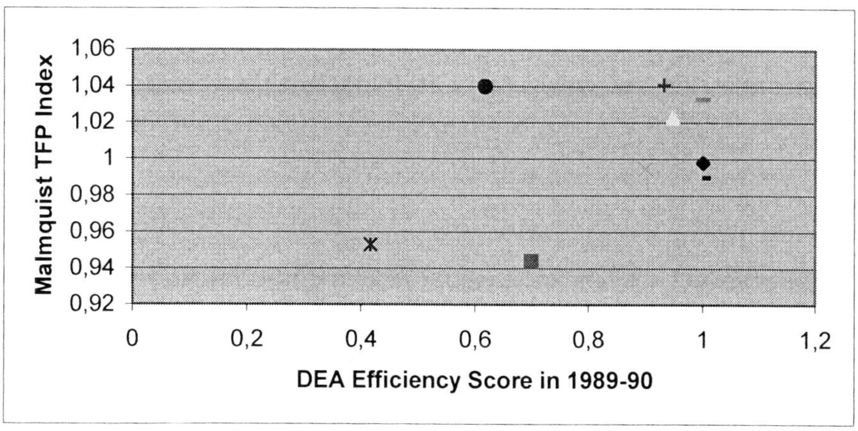

Figure 5. Malmquist Indices Vis-à-vis DEA Scores of 1989-90

9 Some of the statistical analysis carried out has been similar to that of Luoma et al (2000)

Table 3 below presents the statistics related to the DEA scores for each year of the sample period. One notes a fall in mean and median levels of TE scores after the first year viz. 1989-90. Thereafter, the same are found to remain mostly fixed. It must be mentioned in this context that the efficiency scores of the Northeast Frontier Zone, which shows far lower TE scores as compared to other zones, influence the mean values. The average deviation from the mean TE scores, as given by the standard deviations also do not change by much. Also an interesting finding is that virtually throughout the sample period, two Zones, viz. Central and Southeastern Zones were always found to be on the frontier thereby recording a TE score of 1 virtually throughout the sample period. It is also noted that in general, Zones seem to be far away from the respective annual production frontiers, thereby indicating that Zones of IR are possibly not using their inputs optimally.

Table 3. Statistics of DEA Efficiency Scores for Each Year of Sample Period

Year	1990	1991	1992	1993	1994	1995	1996	1997	1998	1999	2000
Mean TE	0.834	0.771	0.792	0.772	0.746	0.721	0.743	0.731	0.749	0.762	0.779
Median	0.931	0.795	0.877	0.807	0.777	0.672	0.71	0.7	0.737	0.76	0.822
s	0.209	0.206	0.223	0.220	0.246	0.236	0.232	0.236	0.237	0.232	0.226
E=1	3	2	2	2	2	2	2	2	2	2	2

Figure 6 tries to study the DEA scores of Zones of IR in 1999-2000 vis-à-vis firm size. Firm size is measured as sum of PKM and TKM (in billion units)[10]. It is observed that there is a strong positive correlation of 0.75 between firm size and efficiency, which is also statistically significant. Thus, one feels inclined to question the rationale behind splitting the existing Zones of IR into smaller Zones. Since these zones have retained the same organizational structure as the ones existing prior to their subdivision, this kind of splitting is devoid of empirical justification.

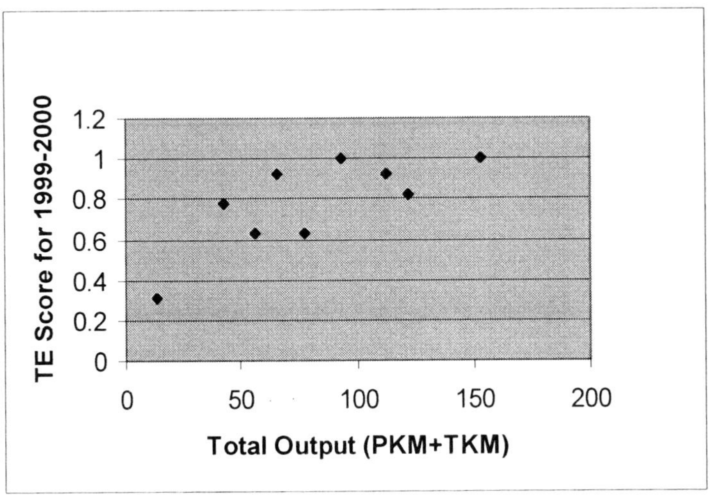

Figure 6. Total Output of IR Vs TE Scores for 1999-2000

[10] This measure has been used in many World Bank studies (Thompson,2000)

4. Policy Recommendations

From the above analysis, some interesting results emerge. It is seen that IR as an organization, has not followed cost minimizing techniques over the years. However, one also notices that managers of Zones of IR have attempted to do their best with the resources available to them, both qualitatively as well as quantitatively. In the current booming Indian economic scenario, wherein a major thrust has been given to infrastructure development, IR will increasingly face competition from other modes. The intense competition, which it is facing from road transport in the freight sector, is likely to spill over to the passenger sector as well. IR is also currently facing competition from the low priced airlines. So it is imperative that in such a scenario, the top management of IR incorporate best practices in production, which would facilitate cost minimization and hence efficiency in production.

The analysis of Technical efficiency scores indicates a positive correlation between firm size and efficiency. Thus, the decision by GOI as well as IR to split the current zones of IR into smaller units does not seem to have any empirical justification, particularly if the smaller zones have the same organizational structure as the existing zones. Moreover, there has not been decentralization in decision making of any kind, which further nullifies the decision to divide the zones into smaller units.

This study has attempted to compute changes in TFP using the Malmquist DEA approach. Two outputs, viz, PKM and TKM were chosen and three inputs, viz. labour, fuel and capital were chosen. The DEA approach is preferable to the cost function approach in that there is no need to adopt the heroic assumption of efficient production. The TFP score has been decomposed into two effects, viz. technical efficiency change or "catch up" effect and shift in the production frontier itself over time.

It is possible to identify some limitations of the study. First of all, the specification of capital output as the real value of investment could be improved upon. It may be preferable to choose track kilometres or rolling stock as the unit of capital. Also, this analysis does not capture the heterogeneity in the outputs. For instance, passenger traffic comprises of sleeper class, second class, three tier AC class, two tier AC class, etc while the freight sector comprises of bulk and non bulk items, petroleum products, agricultural products, gems and jewellery, etc. also, the causes of inefficiency has not been captured in this analysis. This could have been done by regressing the DEA scores of say, a particular year, on various physical parameters to see which of these has a significant influence on efficiency. These could be identified as areas for further research

References

Anand, Y.P.(1998): Reforming Indian Railways' Financial Management in K.B. Verma (ed.) *Readings in Indian Railway Finance*, Academic Foundation, New Delhi.

Banker, R.D, Charnes,A. and Cooper,W.W (1984): Some Models for Estimating Technical and Scale Inefficiencies in Data Envelopment Analysis, *Management Science*,**30**, pp.1978-1092

Caves,D.W, Christensen, L.R. and Diewert, W.E. (1982): The Economic Theory of Index Numbers and the Measurement of Input, output and Productivity, *Econometrica*, Nov, 50, pp1393-1414.

Charnes,A, Cooper,W.W, and Rhodes, E (1978): Measuring the Efficiency of Decision Making Units, *European Journal of Operational Research*,**2,** pp.429-44.

Coelli, T, Estache, A, Perelman,S and Trujillo, L (2003): *A Primer on Efficiency Measurement for Utilities and Transport Regulators*, The World Bank.

Coelli, T, Rao, P and Battese, G. (1997): *An Introduction to Efficiency and Productivity Analysis*, Kluwer Academic Publishers, London.

Coelli,T (1996): *A Guide to DEAP Version 2.1: A Data Envelopment Analysis Computer Program*, Centre for Efficiency and Productivity Analysis, Department of Econometrics, University of New England, Armidale, NSW, 2351, Australia.

GOI (1999): *Ninth Five-Year Plan, Vols. I & II, Planning Commission, Government of India*

Färe,R, Grosskopf, S, and Lovell,C.A.K (1994), *Production Frontiers*, Cambridge University Press, Cambridge.

Jha, R, and Sahni, B, *Industrial Efficiency, an Indian Perspective*, Wiley Eastern Limited

Luoma,K and Järiviö, M (2000), Productivity Changes in Finnish Health Centres in 1988-1995: A Malmquist Index Approach, *VATT Discussion Papers*, Helsinki

Ramanathan, R (2003), *An Introduction to Data Envelopment Analysis*, Sage Publications India Pvt. Limited, New Delhi

Sriraman,S (2002), Indian Railway Finances- A Critical Review of Basic Issues and Emerging Options, *Journal of Indian School of Political Economy*, pp.377-384

Thanasoulis, E. (2001): *Introduction to the Theory and Application of Data Envelopment Analysis*, Kluwer Academic Publishers.

Thompson, L.(2000): *Restructuring Indian Railways*, PowerPoint presentation available at website http://www.worldbank.org/transport/rail/india_00.pdf

In: Economics of Technological Change
Editor: Thomas L. Beane, pp. 123-143

ISBN: 978-1-60021-237-6
© 2007 Nova Science Publishers, Inc.

Chapter 7

TECHNOLOGICAL COLLABORATION AND COLLUSION: A TRIGGER STRATEGY ANALYSIS

Gamal Atallah
Department of Economics, University of Ottawa
and *CIRANO*

ABSTRACT

The paper analyses the relationship between technological collaboration and collusion. Firms can collude or defect on the output market, but with the novelty that there is a possibility of developing a new technology jointly. The development of the new technology is conditional on prior collusion by firms. The paper derives three main results. First, it is found that technological collaboration facilitates product market collusion, but only in the short-run, by creating an equilibrium where firms collude initially in order to develop the new technology, and then defect afterwards. Nonetheless, there is less defection compared with the standard trigger strategy model. Second, it is shown how the equilibrium depends on the discount factor; in particular, delayed defection is sustained for intermediate ranges of the discount factor. Finally, the comparative statics analysis shows how the equilibrium is affected by changes in the environment. An increase in the scope of the market, an increase in the number of firms, and a lower quality technology all contribute to making collusion more difficult. Whereas permanent collusion is affected only by changes in the number of firms, delayed defection is affected also by the scope of the market and the quality of the new technology. The results are shown to extend to Bertrand competition, although in the latter case there is less delayed defection than under Cournot competition.

1. INTRODUCTION

What is the relationship between technological collaboration and collusion? Are firms who collaborate together in the development of new products and processes more likely to collude on the output market? This question lies at the heart of antitrust policy dealing with high-tech industries. As Vonortas notes: "frequent meetings to conduct R&D means that the technological interests of two firms overlap to a large extent, which may open an effective avenue for punitive action against a defector: shunning from a technological area(s) or from a technical standard over a period of time." (Vonortas, 2000:247). This paper addresses this question through a model where firms can collude or defect on the output market, but with the novelty that there is a possibility of developing a new technology jointly.

The current model endogenizes both output collusion and technological collaboration. In a Cournot setting, firms compete for an infinite number of periods. In the first period, firms produce using the old technology. After the first period, firms have the possibility of forming a Research Joint Venture (RJV) to develop a cost-reducing technology. However, firms will work jointly on developing the new technology if and only if no defection has taken place prior to that period. Moreover, firms follow the generalized trigger strategy that they will collude forever if and only if no defection has taken place in the past. Hence there are three possible equilibria: defection in all periods, delayed defection (i.e. temporary collusion, which implies defection after the development of the new technology), and permanent collusion.

It is found that the equilibrium strategy depends on the extent to which firms discount the future. If firms are very patient, permanent collusion will be sustained. If firms are moderately patient, they will collude prior to the development of the new technology, but will defect afterwards. And if firms are not very patient, they will defect early on, so that the RJV will not be formed. Moreover, compared with the standard model where there is no new technology, the prospect of technological cooperation facilitates product market collusion, but only in the short-run. That is, the extent of permanent collusion is not affected; the only change is that there is a range of discount factors such that the equilibrium strategy changes from defection to delayed defection, because of the presence of the new technology.

We also analyze the impact of the parameters of the model on the equilibrium. A better technology developed through technological collaboration reduces defection and increases delayed defection, but has no effect on permanent collusion. An increase in the number of firms increases defection, and reduces short-run and total collusion. An increase in the scope of the market (an increase in demand or a reduction in initial production costs) increases defection and reduces delayed defection. The explanation for these results lies in the effect of changes in the environment on defection profits and collusive profits. Defection profits increase more than collusive profits in response to an increase in the scope of the market. Similarly, defection profits decline less following an increase in the number of firms than do collusive profits. Finally, qualitatively similar results are obtained under Bertrand competition; however, there is less delayed defection under Bertrand competition than under Cournot competition.

There is a small literature on the relationship between R&D collaboration and collusion. Brod and Shivakumar (1997) analyze R&D cooperation with and without output cooperation, with both R&D cooperation and collusion considered as exogenous. Martin (1995) develops a theoretical model and finds that R&D cooperation increases the likelihood of product market collusion. Cabral (2000) studies R&D collaboration under price competition and focuses on the interaction between R&D levels and market prices. Lambertini et al. (2002) focus on the impact of product differentiation on the relationship between R&D collaboration and collusion. van Wegberg (1995) analyzes the interaction between R&D alliances, mergers and collusion. Suetens (2003) runs an experiment analyzing the impact of R&D cooperation on prices. Lin (1998) finds that firms behave less aggressively in the product market in order to soften R&D competition in a subsequent R&D race, and that R&D cooperation eliminates this collusive conduct.[1] A common finding of the literature (with the exception of Cabral) is that R&D cooperation facilitates product market collusion.

Models such as Brod and Shivakumar (1997) assume that both output collusion and technological collaboration are exogenous, and compare the profits and social welfare associated with them. However, it is not sufficient for collusion and/or technological collaboration to be privately and/or socially profitable for them to occur.

This paper contributes to the literature on the relationship between technological collaboration and collusion in three ways. First, it shows that technological collaboration facilitates product market collusion, but only in the short-run. As will be seen, the temporary nature of collusion arising from technological collaboration arises from the temporary nature of research alliances themselves. Second, it shows how the equilibrium strategy depends on the discount factor. Third, it studies the impact of the number of firms, the size of the market, and the quality of the new technology on the equilibrium.

The paper is organized as follows. The model is solved and the equilibrium strategies derived in the next section. Section 3 analyzes the impact of the quality of the technology, the number of firms, and the scope of the market on the equilibrium. In section 4, the results are derived under price competition, and are shown to be qualitatively similar to those obtained under quantity competition. The last section concludes.

2. The Model

There are n identical firms producing a homogeneous product and competing à la Cournot. Firms face the linear inverse demand $p(Y)=a-Y$, where Y denotes total industry output. The output of firm i is denoted y_i. Initially, the marginal cost of production is $c_I=c$. Firms have the possibility of forming a RJV to develop a superior technology, which reduces the marginal cost of production by Δ, resulting in the marginal cost $c_J=c-\Delta$. For simplicity, it is assumed that no firm can develop the technology on its own.[2]

This is an infinite horizon game. In the first period, firms produce using the initial technology. Before engaging in second period production, firms have the possibility of forming the RJV, which would make the new technology available for production starting from the second period and for all subsequent periods. In each subsequent period, firms produce using the initial technology if no RJV has been formed, or using the new technology if an RJV has been formed. In each period, firms can collude or defect on the output market.

More specifically, on the output market, firms can take three possible actions: collude (c), defect (d), or play Nash (N). Collusion means joint profit maximization, while defection means maximizing own-profits while the other firms think that collusion is in place. Nash is the equilibrium where firms would end when collusion is no longer sustained (Cournot competition).

We let π_T^m denote the per-period profits of a firm, with $T\in\{I,J\}$, and $m\in\{c,d,N\}$. For any technology, $\pi^d>\pi^c>\pi^N$. Similarly, because the new technology is superior, for any market

---

1 A more remote literature deals with the relationship between patent protection and collusion. Kultti et al. (2004) argue that patenting makes collusion more difficult, since it protects the deviating firm against retaliation, and because the deviating firm is uncertain whether it will get the patent.

2 The specific activity of the RJV is not specified. It can include sharing information, coordinating R&D expenditures, setting a joint R&D lab, etc.

structure, $\pi_J > \pi_I$.[3] The per-period profit of firm i under the initial technology is $\pi_i = (p-c)y_i$. If the new technology is being used, the per-period profit is given by $\pi_i = (p-c+\Delta)y_i$. Given that all variables are time invariant (except for the possible change of technology), variables are not indexed by time, for simplicity.

For a given technology, profits depend on the prevailing regime. Using technology $T \in \{I, J\}$, the price, per firm output, and profits of playing Nash (Cournot competition) are:

$$p_T^N = \frac{a + nc_T}{n+1}, \quad y_T^N = \frac{a - c_T}{n+1}, \quad \pi_T^N = \left(\frac{a - c_T}{n+1}\right)^2 \tag{1}$$

The equilibrium under collusion is

$$p_T^C = \frac{a + c_T}{2}, \quad y_T^C = \frac{a - c_T}{2n}, \quad \pi_T^c = \frac{(a - c_T)^2}{4n} \tag{2}$$

And defection would entail (for the defecting firm)

$$p_T^d = \frac{a(n+1) + c_T(3n-1)}{4n}, \quad y_T^d = \frac{(a - c_T)(n+1)}{4n}, \quad \pi_T^d = \left(\frac{(a - c_T)(n+1)}{4n}\right)^2 \tag{3}$$

The focus of the model is on the interaction between the decision to respect the cartel agreement (maximizing joint profits), and the decision to form the RJV. Therefore, it is assumed that for firms to agree to develop the new technology, collusion by all firms must be observed in the first period. Otherwise, if a firm defects in the first period, the other firms will refuse to work on the development of the new technology. If a firm defects early on, the other firms will punish it not only through defection on the output market, but also through refusing to develop the new technology jointly. This can be seen as an enriched trigger strategy, where punishment is in terms of both output levels and technology development. If a firm defects, it benefits during the defection period (because $\pi^d > \pi^c$) but loses afterwards (because $\pi^c > \pi^N$).[4]

In sum, there are three possible strategies:[5]

-*TC* (*Total* collusion): Collude in the first period, develop the new technology, and collude in each subsequent period using the new technology.

-*DD* (*Delayed defection*): Collude in the first period, develop the new technology, and defect in the second period, which will trigger Nash behavior forever, using the new technology.

---

3 While the cost of forming the RJV is not written explicitly, the assumption $\pi_J > \pi_I$ means that the benefits of forming the RJV outweigh its costs.

4 Naturally, the formation of an RJV need not be associated with product market collusion. Rather, here we consider a market where firms are willing to collude, and ask the question, how is this willingness affected by the possibility of technological cooperation.

5 Because this is an infinite horizon game, many other equilibria exist. Here we focus on the trigger strategy equilibrium involving Nash behavior and collusion.

-*TD* (*Total defection*): Defect in the first period, which will trigger Nash behavior forever, using the old technology.

Let δ represent the discount factor. The profits of each strategy can be written as follows:

$$\Pi^{TC} = \pi_I^c + \delta\pi_J^c + \delta^2\pi_J^c + \ldots = \pi_I^c + \frac{\delta}{1-\delta}\pi_J^c$$

$$\Pi^{DD} = \pi_I^c + \delta\pi_J^d + \delta^2\pi_J^N + \delta^3\pi_J^N + \ldots = \pi_I^c + \delta\pi_J^d + \frac{\delta^2}{1-\delta}\pi_J^N \tag{4}$$

$$\Pi^{TD} = \pi_I^d + \delta\pi_I^N + \delta^2\pi_I^N + \ldots = \pi_I^d + \frac{\delta}{1-\delta}\pi_I^N.$$

2.1. The Equilibrium

It is not possible to perform all the analysis of the results analytically, hence some results will be derived by attributing numerical values to some of the parameters of the model.[6]

Proposition 1. *The equilibrium will be:*

- *Total collusion: firms will collude in all periods iff* $\delta > \delta^{**}$.
- *Total defection: firms will defect in all periods iff* $\delta < \delta^{*}$.
- *Delayed defection: firms will collude in the first period, and defect in all subsequent periods, iff* $\delta \in [\delta^{*}, \delta^{**}]$,

with δ^{*} *and* δ^{**} *given by* (6) *and* (5), *such that* $\delta^{*} < \delta^{**}$.

Proof.

$$\Pi^{TC} > \Pi^{DD} \quad iff \quad \delta > \delta^{**} \equiv \frac{\pi_J^d - \pi_J^c}{\pi_J^d - \pi_J^N} = \frac{(n+1)^2}{n^2 + 6n + 1} \tag{5}$$

To determine whether the new technology will be developed in the first place, we need to compare the costs and benefits of defection in the first period, before the possibility of technological collaboration is available.

$$\Pi^{DD} > \Pi^{TD} \quad iff \quad \delta > \delta^{*} \equiv \frac{\pi_J^d + \pi_I^d - \pi_I^c - \pi_I^N - \sqrt{(\pi_I^c - \pi_I^d + \pi_I^N - \pi_J^d)^2 + 4(\pi_I^c - \pi_I^d)(\pi_J^d - \pi_J^N)}}{2(\pi_J^d - \pi_J^N)} \tag{6}$$

It remains to show that $\delta^{**} > \delta^{*}$. As this is not feasible analytically, we will show that it is true for different combinations of parameter values.

6 While many of the results of the Cournot model are derived using numerical simulations, the results of the Bertrand model in section 4 are all derived analytically, and confirm the results obtained here.

Table 1. Comparison between δ^* and δ^{**} for different parameter values

a	n	c	Δ	δ^*	δ^{**}
1000	2	50	25	0.384	0.529
2000	2	50	25	0.441	0.529
1000	5	50	25	0.579	0.643
1000	2	55	25	0.383	0.529
1000	2	50	20	0.404	0.529
1000	20	30	5	0.836	0.847

In all cases $\delta^{**}>\delta^*$, meaning that there exists a range of discount factors such that firms engage in *DD*, and that for higher values of the discount factor firms engage in *TC*. Many other combinations have been tried, all confirming that $\delta^{**}>\delta^*$. ■

(5) is the equivalent of the standard collusion condition for the trigger strategy, indexed by the technology subscript *J*. This condition means that, given that the new technology has already been developed, firms will collude if they are patient enough.

(6) says that firms will prefer to collude in the short-run (the short-run here means the first period) in order to benefit from the new technology if the discount rate is high enough. After the development of the new technology, collusion cannot be sustained. If, on the other hand, $\delta<\delta^*$, the new technology will not be developed, and Nash behavior will be observed forever.

The following figure illustrates how the equilibrium is determined by the discount factor:

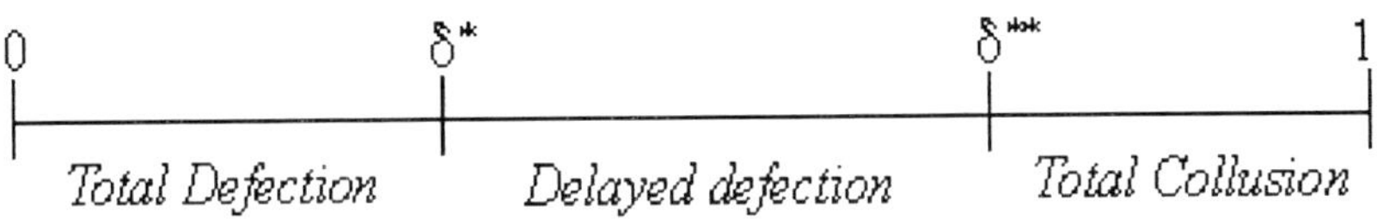

Figure 1. The discount factor and collusion regimes

When the discount factor is very low, the possibility of technological collaboration is not sufficient to sustain collusion, and firms defect anyway. For intermediate levels of discounting, firms collude in the first period to ensure technological collaboration. However, once the new technology has been developed, the incentives for collusion collapse, and Nash behavior prevails. Finally, when firms are sufficiently patient, they collude both before and after the development of the new technology. In this case it is likely that firms would have colluded anyway, and technology would have had no effect on the equilibrium.

With *TD* and *TC*, firms either defect or collude. However, under *DD*, we observe something that is absent from the traditional trigger strategy model and from the literature: delayed defection. Firms collude to reap the benefits of technological collaboration, and then defect. In this sense, the delayed defection we observe for intermediate levels of δ can be said to be "technology-induced", since its sole purpose is to acquire the new technology. From this point of view, it can be said that technological collaboration can facilitate (temporary) collusion.

Above numerical simulations were used to show that $\delta^{**}>\delta^{*}$ because of the analytical complexity of δ^{*}. It is possible to provide an analytical proof of this inequality for the simplified case where $p(Y)=1-Y$ and $\Delta=c$ (this means that the new technology drives marginal production costs to zero). In this simplified model δ^* depends only on c and n, although it is still an analytically cumbersome function. Moreover, an interior solution for p and Y requires $c\in(0,1)$. Without loss of generality, we can plot $\delta^{**}-\delta^{*}$ in the (cxn) space (see figure 2). As this figure shows, the resulting curve is always positive, proving the result. Moreover, plotting the curve for higher values of n would not change that result.

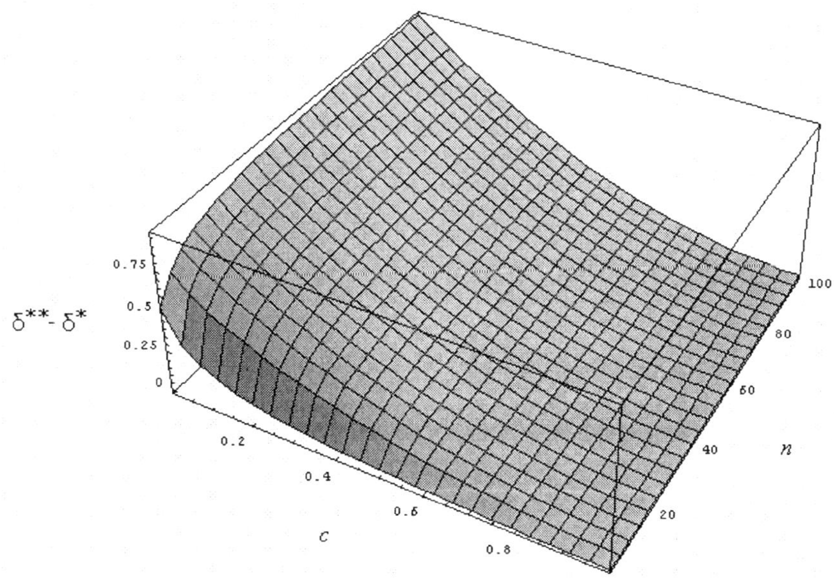

Figure 2. $\delta^{**}-\delta^{*}$ in the simplified Cournot model. $D(p)=1-p$, $c_I=c$, $c_J=0$

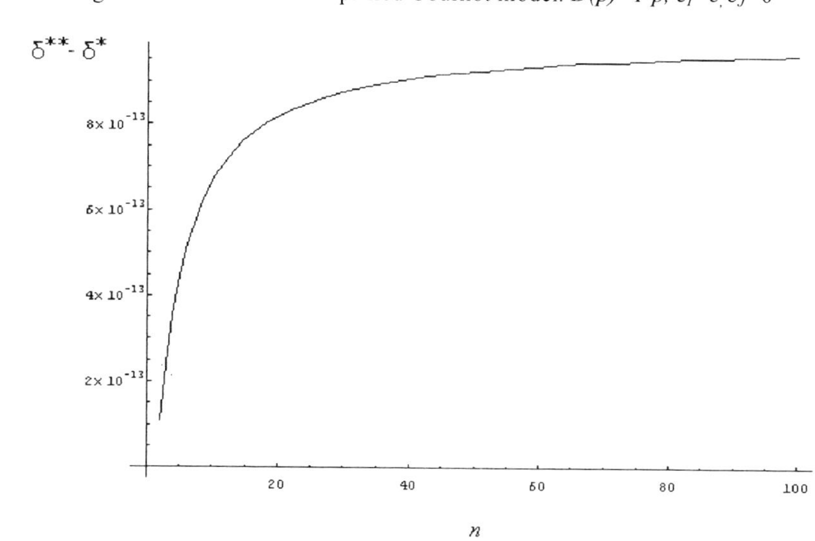

Figure 3. $\delta^{**}-\delta^{*}$ in the simplified Cournot model($c\rightarrow 1$)

In addition, note that δ**-δ* declines with c (we know that from table 1 and from figure 2). Evaluating δ**-δ* in the limit case where $c \to 1$ yields a function which depends only on n, which is plotted in figure 3. As the figure shows, the difference is clearly positive.

The model allows us to link the path of prices with the collusion regime chosen by firms. From equations (1) through (3), we know that $p_I^C > p_J^C > p_J^N$. The comparison between p_J^C and p_I^N is ambiguous. We have that $p_J^C > p_I^N$ iff

$$n > \frac{a+c+\Delta}{a+c-\Delta} \tag{7}$$

i.e., n is high and Δ is low. When (7) is satisfied, we have that

$$p_I^c \quad > \quad p_J^c \quad > p_I^N \quad > \quad p_J^N$$

$$\frac{a+c}{2} > \frac{a+c-\Delta}{2} > \frac{a+c}{n+1} > \frac{a+c-\Delta}{n+1} \tag{8}$$

When, however, (7) is not satisfied, we have that

$$p_I^c \quad > \quad p_I^N \quad > \quad p_J^C \quad > \quad p_J^N$$

$$\frac{a+c}{2} > \frac{a+c}{n+1} > \frac{a+c-\Delta}{2} > \frac{a+c-\Delta}{n+1} \tag{9}$$

Assume to focus ideas that firms were not colluding prior to starting working on the joint development of the new technology. Figure 4 illustrates the price paths based on (8), where concentration is low and the technology is minor. Figure 5 shows the price paths based on (9), which is more likely to obtain when concentration is high and the technology is major. On both figures, the development of the technology is always associated with a price hike, followed by a decline, whether the regime is *TC* or *DD*. *TC* is associated with a slight decline after the new technology is introduced, due to lower production costs. The price decline is stronger under *DD*, due to the combined effect of lower production costs and the breakdown of collusion. The ultimate price is always below *TD* for *DD*, but may be above or below *TD* for *TC*, depending on n and Δ.

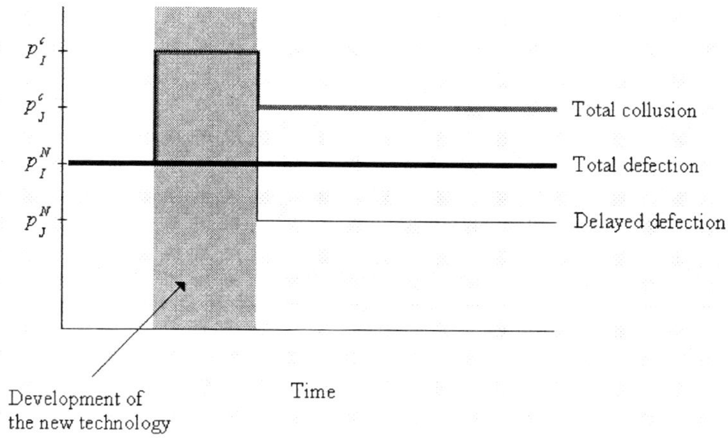

Figure 4. Price paths. High n, low Δ.

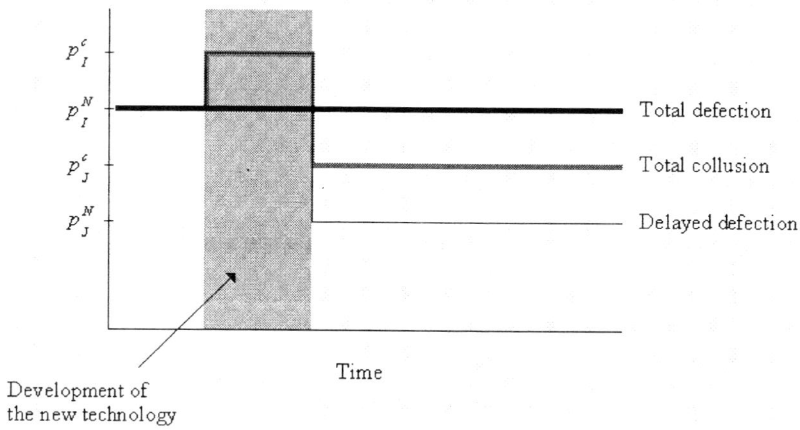

Figure 4. Price paths. Low n, high Δ.

2.2. Comparison with the Standard Model

How does the extent of collusion in this model compare with the standard model, where technology is not an issue? To answer that question, we first derive the equilibrium of the standard model. Let δ_S^{**} denote the critical discount factor which separates collusion from defection in the standard model. In this case firms would be using the initial technology (*I*) under both defection and cooperation. Moreover, the only possible strategies are *TD* and *TC*. The critical discount factor is

$$\delta_s^{**} = \frac{\pi_I^d - \pi_I^c}{\pi_I^d - \pi_I^N} = \frac{(n+1)^2}{n^2 + 6n + 1} \tag{10}$$

Note that $\delta_s^{**}=\delta^{**}$: the regions of *TC* coincide perfectly. What differs between the two models is that in the model with *DD*, there is a range to the left of δ^{**} where *DD* is sustainable. This range exists solely because of the prospect of technological collaboration.

Proposition 2. *The existence of the possibility of technological collaboration reduces the range of discount factors such that defection is the equilibrium.*

Proof. In the standard model, defection occurs when $\delta<\delta_s^{**}$. In the model with technology cooperation, defection occurs when $\delta<\delta^*$. And we have that $\delta^*<\delta^{**}=\delta_s^{**}$. ■

3. Comparative Statics

We now wish to analyze how the likelihood of observing each regime is affected by the number of firms, the cost advantage of the new technology, and the scope of the market. Given that differentiating δ^* analytically w.r.t. the parameters of the model is difficult, we rely on numerical simulations. To derive the desired comparative statics, we fix all other model parameters, and plot δ^* as a function of the relevant parameter. The following numerical parameterization is considered as the benchmark case: a=1000, c=50, Δ=5, n=2.[7]

Notice that δ^{**} depends only on n, whereas δ^* depends on a, n, c and Δ. This means that the determinants of short-run and long-run collusion are different: the latter depends only on the number of firms, whereas the former depends also on the quality of the new technology and on the scope of the market.

3.1. Quality of the New Technology

Consider first the effect of the quality of the new technology on the equilibrium.

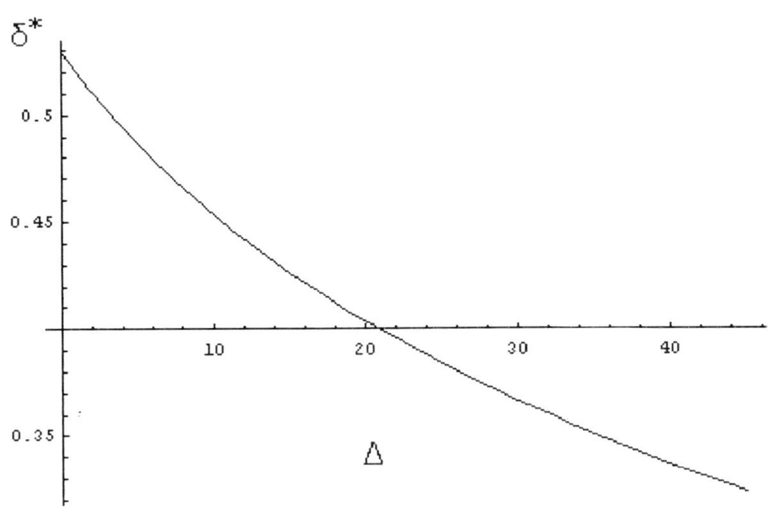

Figure 6. Effect of Δ on δ*.

7 None of the results that follow depend on this parameterization.

Proposition 3. *An increase in the quality of the new technology* (Δ) *reduces defection and increases delayed defection, but has no effect on total collusion.*

Proof. Figure 6 shows that δ^* decreases with Δ. And from (5), δ^{**} is independent of Δ. ■

By shifting δ^* to the left, the increase in Δ increases the range of discount factors such that firms choose *DD*, and reduces the scope of *TD*. The shift in Δ increases the profitability of the new technology relative to the old technology, for two reasons: the marginal production cost is reduced further for each unit produced; and given that a lower cost means increased output for all firms, profits are increased through the production of more output. This makes it worthwhile to wait until the new technology has been developed to defect.

Π^{TD} is not affected by the improvement in the quality of the technology (because it is never developed nor used), while Π^{DD} increases with Δ, because the profits of the second period and onward are higher due to the lower marginal cost. Hence, the tradeoff tilts toward *DD*.

3.2. The Number of Firms

We now ask how the number of firms affects the incentives for collusion and cooperation.

Proposition 4. *An increase in the number of firms* (n) *increases defection, and reduces short-run* (*except when n is very high*) *and total collusion.*

Proof.

$$\frac{\partial\delta^{**}}{\partial n} = \frac{4(n^2-1)}{(n^2+6n+1)^2} > 0 \tag{11}$$

As for δ^*, figure 7 shows that it increases with n. Therefore there is more *TD* and also less *TC*. To determine the effect on *DD*, we need to analyze what happens to the difference δ^{**}-δ^*. We plot this difference as a function of n in figure 8:

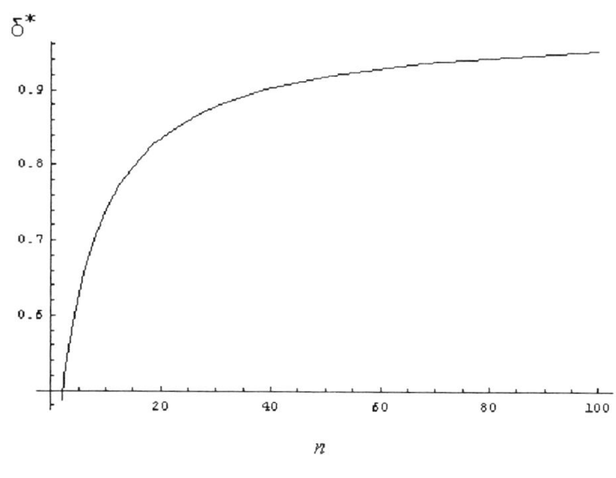

Figure 7. Effect of n on δ*.

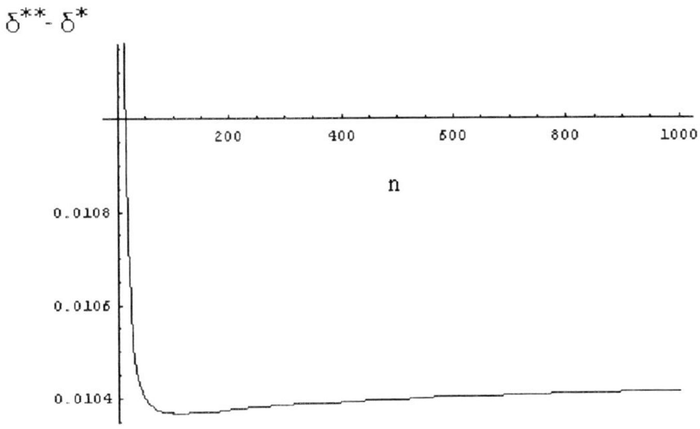

Figure 8. Effect of n on δ^{**}-δ^{*}.

We see that the difference is decreasing in n for up to n above 100, meaning that the range of discount factors such that *DD* is the equilibrium is *squeezed* by the increase in n. For higher values of n, δ^{**}-δ^{*} increases slightly with n, indicating an increase in *DD*. ■

The effect of n on the equilibrium strategies can be understood by looking at the components of Π^{TC}, Π^{DD} and Π^{TD}. The shift to the right of both δ^{*} and δ^{**} means that there can be up to three regions (see figure 9) where the increase in n changes the equilibrium strategy:

-from *DD* to *TD* (region 1);
-from *TC* to *TD* (region 2); and
-from *TC* to *DD* (region 3).

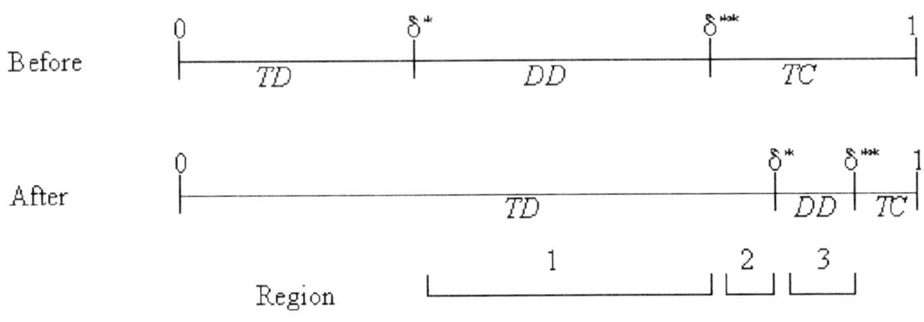

Figure 9. Effect of competition on the equilibrium.

The explanation of each of these changes lies in how n affects profits. Consider first region 1. The comparison here is between the effect of n on Π^{DD} and Π^{TD}. Given that at the left boundary of this region it was the case that $\Pi^{DD}=\Pi^{TD}$, it must be that an increase in n reduces Π^{TD} less than it reduces Π^{DD}. To understand the reason for this, we look at the components of total profits in each case. For any strategy, profits can be divided into three groupings: period 1, period 2, and the periods following period 2. For Π^{DD}, in the first period profits are π_I^c, in the second period they are π_J^d, and in all subsequent periods they are π_J^N (see (4)). Similarly, for Π^{TD}, profits can be decomposed into π_I^d in the first period, π_I^N in the second period, and π_I^N in all subsequent periods. We compare how the profits in each period are affected by the increase in n. In the first period, the decline in profits under *DD* is more severe than under *TD*, because

$$\left|\frac{\partial \pi_i^d}{\partial n}\right| - \left|\frac{\partial \pi_i^c}{\partial n}\right| = -\frac{(a-c)^2(n-1)}{8n^3} < 0 \tag{12}$$

Similarly, the decline in the third and following periods under *DD* is more severe, because

$$\left|\frac{\partial \pi_i^N}{\partial n}\right| - \left|\frac{\partial \pi_J^c}{\partial n}\right| = -\frac{2\Delta[2(a-c)+\Delta]}{(n+1)^3} < 0 \tag{13}$$

Even though the decline of second period profits under *TD* may be more or less severe than the decline in profits under *DD*, the effects described in (12) and (13) always dominate. Hence the reason why there is a parameter region where the equilibrium strategy becomes *TD* instead of *DD* is that the decline in defection profits is lower than the decline in collusive profits, and that the decline in Nash profits (using the old technology) is lower than the decline in collusive profits (using the new technology).

In region 2, the equilibrium strategy becomes *TD* instead of *TC*. And this is even though before the increase in n, it was the case that $\Pi^{TC} > \Pi^{TD}$ (because we were to the right of δ^{**}; see figure 9). Therefore, the decline in Π^{TC} must be not only larger than the decline in Π^{TD}, but must also be substantial enough to overcome that initial advantage. Looking at the components of the profits of each strategy (4), we see that for Π^{TD}, profits can be decomposed into π_I^d in the first period, π_I^N in the second period, and π_I^N in all subsequent periods. Similarly, for Π^{TC}, profits can be decomposed into π_I^c in the first period and π_J^c for all subsequent periods. We wish to determine what's behind the sharp decline in Π^{TC}. From (12), we know that the decline in collusive profits in the first period under *TC* is larger than the decline in defection profits under *TD* in that same period. This effect favors *TD*. In all subsequent periods, the decline in Nash profits under *TD* may be smaller or larger than the decline in collusive profits under *TC*, but irrespective of the sign of this difference, the first effect dominates, and Π^{TC} declines more than Π^{TD}.

Finally, in region 3, the equilibrium becomes *DD* instead of *TC*. The components of Π^{TC} are π_I^c in the first period and π_J^c for all subsequent periods. The components of Π^{DD} are π_I^c in the first period, in the second period they are π_J^d, and in all subsequent periods they are π_J^N. Because the first period profits are the same, the result will hinge upon what happens in subsequent periods. In the second period, the decline in π_J^c under *TC* is larger than the decline in π_J^d under *DD*, that is,

$$\left|\frac{\partial \pi_J^c}{\partial n}\right| - \left|\frac{\partial \pi_J^d}{\partial n}\right| = \frac{(a-c+\Delta)^2(n-1)}{8n^3} > 0 \tag{14}$$

This favors *DD*. However, the decline in π_J^N under *DD* in the third period may be smaller or larger than the decline in π_J^c under *TC*. But in all cases the effect of the second period dominates, and more *DD* is induced at the expense of *TC*.

Regions 1 and 3 always exist, but whether region 2 (the region where the equilibrium goes from *TC* to *TD*) exists depends on the positions of δ^{**} before the increase in n and of δ^* after the increase in n. Let $\delta^{**}(n)$ denote the old value of δ^{**} before the increase in the number of firms, and let $\delta^*(n+z)$ denote the value of δ^* after the increase in the number of firms, where z denotes the number of firms that enter the market. For region 2 to exist, it must be that $\delta^*(n+z)>\delta^{**}(n)$. Numerical simulations show that this is more likely to be the case when *a*) n is low, or *b*) n and z are high. Hence, either the industry is initially concentrated, or the industry is not concentrated but entry is massive. For instance, with $z=1$ (one firm enters), region 2 will exist iff $n\leq 12$. This is consistent with our analysis: a low n shifts $\delta^{**}(n)$ to the left, while a high z shifts $\delta^*(n+z)$ substantially to the right, making the existence of region 2 more likely. The significance of region 2 lies in that the equilibrium changes sharply, from *TC* to *TD*, without going through the intermediate phase of *DD*.

For a large interval of n, δ^{**}-δ^* is decreasing (sharply) in n, meaning that there is less *DD*. Only as n becomes very large does the difference increase slightly with n. However, at such large values of n collusion is unlikely to hold, and unlikely to involve all firms in the industry. Therefore, we consider that the dominant effect of the increase in n is to decrease *DD*.

To summarize, the equilibrium goes from *DD* to *TD* (region 1) because the decline in defection profits is smaller than the decline in collusive profits, and because the decline in Nash profits is smaller when firms are using the old technology than when they are using the new technology. The equilibrium goes from *TC* to *TD* (region 2) because the decline in defection profits is lower than the decline in collusive profits. Finally, the equilibrium goes from *TC* to *DD* (region 3) because the decline in defection profits is lower than the decline in collusive profits.

The fact that the increase in n makes *TC* harder is well known, as the cartel profits have to be shared among a larger number of firms, which threatens the internal stability of the cartel. Here it is shown that a similar result holds for *DD*. The gains from collusion are reduced by competition whether the old or the new technology is used. As firms have to sacrifice short-run profits (because they could defect in the first period and make $\pi^d>\pi^c$) to obtain the technology, the increase in n reduces the incentives to make this sacrifice, increasing the incentives for *TD*.

Moreover, the shift of δ^{**} to the right means that there are situations where firms would engage in *TC*, but because of the increase in n now decide to engage in *DD* only. This is because the increase in n makes it still worthwhile to sacrifice defection profits in the first period to acquire the technology, but makes it unprofitable to collude forever.

A common factor in all three regions is that the decline in defection profits is lower than the decline in collusive profits, both with the old and the new technology. This is due to the effect of n on prices and output under the two scenarios. From equations (1) through (3) we know that $|\partial y^d/\partial n|<|\partial y^c/\partial n|$, so that an increase in n reduces the output of a firm by more if it is colluding than if it defects. This is because the defecting firm does not internalize the impact of its action on the other firms' profits, and hence reduces its output by less. And even though p^d declines with n while p^c is not affected by n, revenues decline by more under collusion (see (12)), implying that collusive profits decline more than defection profits following an increase in n.

3.3. The Size of the Market

We now consider the impact of the size of the market.

Proposition 5. *An increase in demand* (a) *or a reduction in the initial production cost* (c) *increases defection and reduces delayed defection.*

Proof. Figure 10 shows that δ^* increases with a. Figure 11 shows that δ^* decreases with c. And from (5) we know that δ^{**} is independent of a and c. ■

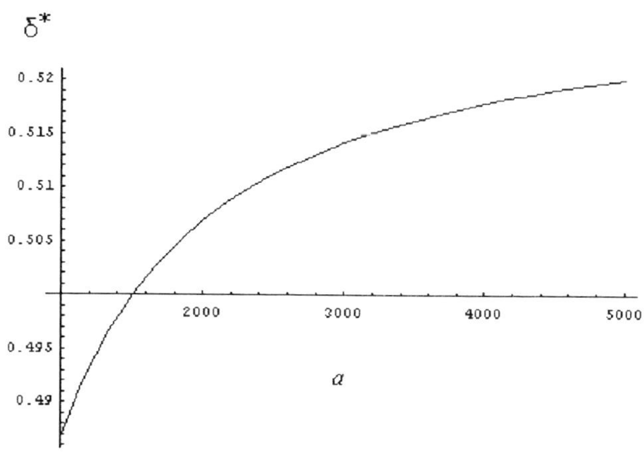

Figure 10. Effect of a on δ^*.

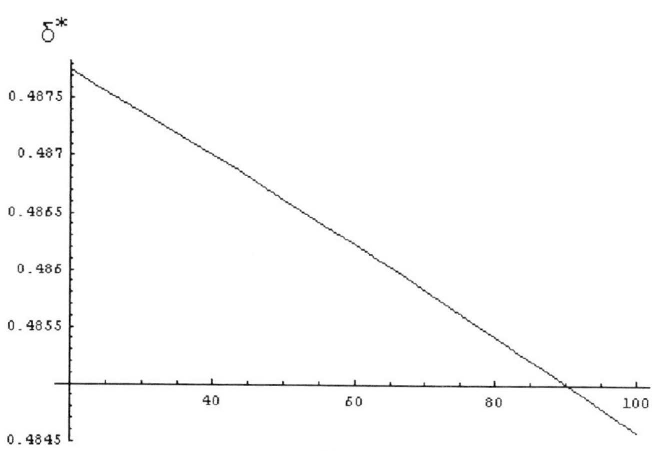

Figure 11. Effect of c on δ^*.

This result can be understood by looking at the components of Π^{DD} and Π^{TD} and how they are affected by the increase in demand. Note that

$$\frac{\partial(\pi_I^d - \pi_I^c)}{\partial a} = \frac{(a-c)(n-1)^2}{8n^2} > 0 \tag{15}$$

that is, the increase in demand increases defection profits more than it increases collusive profits. These are the profits made by a firm in the first period under *TD* and *DD*, respectively. This increases the incentives for defection in the first period. Even though the effects on the other components go in the opposite direction, i.e. increase the components of Π^{DD} more than they increase the components of Π^{TD},[8] the effect of the first period dominates the effect of all other periods (in the neighborhood of δ^*), hence δ^* moves to the right. Said otherwise, the increase in demand increases Π^{TD} more than it increases Π^{DD} in the neighborhood of δ^*. A reduction in *c* has the opposite effect of an increase in *a*.

This result can be explained by looking at the impact of demand on output and prices. From equations (1) through (3), it is easy to see that $\partial y^d/\partial a > \partial y^c/\partial a$, so that the expansion in demand increases the output of the firm more if it is defecting that if it is cooperating. Again, this is because the firm does not internalize the impact of its output expansion on the other firms' profits. And even though $\partial p^d/\partial a < \partial p^c/\partial a$, revenues increase by more under defection, hence defection profits increase by more than collusive profits.

Therefore *DD* is more likely to be observed in markets with low demand, high production costs, high concentration, and when the benefits of the new technology are substantial. These characteristics resemble those of a new market or a new product.[9] As the market matures (demand is high, production costs are low due to learning and experience, competition is well-established, and further technological developments are limited), more defection, and less short-run and total collusion should be observed. This result is also consistent with the prediction of price wars during booms (à la Rotemberg and Saloner, 1986).

4. Delayed Defection in a Bertrand Market

To test the robustness of the results derived above, in this section the trigger strategy is applied in a Bertrand market. Everything is the same as before, except that now firms compete in prices. Using technology $T\in\{I,J\}$, the profits of playing Nash (Bertrand competition) are $\pi_T^N=0$, since price competition drives prices to marginal costs. The profits of collusion are

$$\pi_T^c = \frac{(a-c_T)^2}{4n} \tag{16}$$

the same as under Cournot competition. Finally, the profits of defection are

[8] It is easy to verify that $\partial(\pi_J^d-\pi_I^N)/\partial a>0$ and $\partial(\pi_J^N-\pi_I^N)/\partial a>0$, which favors *DD*.

[9] In a sense, *DD* is difficult to observe, because it is temporary.

$$\pi_T^d = \frac{(a-c_T)^2}{4} \tag{17}$$

since defection implies reaping the entire monopoly profits through undercutting the cartel by charging

$$p_T^d = \frac{a+c_T}{2} - \varepsilon \tag{18}$$

where $(a+c_T)/2$ is the cartel price and $\varepsilon>0$ is an arbitrarily small deviation. Hence, price competition entails greater defection profits, but also a more severe punishment, than quantity competition.

Using the same notation as above, the profits of the three possible strategies are

$$\begin{aligned} \Pi^{TC} &= \pi_I^c + \delta\pi_J^c + \delta^2\pi_J^c + \ldots = \pi_I^c + \tfrac{\delta}{1-\delta}\pi_J^c \\ \Pi^{DD} &= \pi_I^c + \delta\pi_J^d \\ \Pi^{TD} &= \pi_I^d. \end{aligned} \tag{19}$$

Proposition 6. *Under Bertrand competition, the results of proposition* 1 *hold, with*

$$\delta^* = \frac{(a-c)^2(n-1)}{n(a-c+\Delta)^2}, \quad \delta^{**} = \frac{n-1}{n} \tag{20}$$

Proof.

Substituting (16) and (17) into (19), we have that

$$\Pi^{TC} = \frac{(a-c)^2 + \delta\Delta[2(a-c)+\Delta]}{4(1-\delta)n} \tag{21}$$

$$\Pi^{DD} = \frac{(a-c)^2(1+\delta n) + \delta\Delta n[2(a-c)+\Delta]}{4n} \tag{22}$$

$$\Pi^{TD} = \frac{(a-c)^2}{4} \tag{23}$$

from which it follows that $\Pi^{TC}>\Pi^{DD}$ iff $\delta>\delta^{**}$ and $\Pi^{DD}>\Pi^{TD}$ iff $\delta>\delta^*$. Moreover, $\delta^{**}>\delta^*$ since

$$\delta^{**} - \delta^* = \frac{(2(a-c)+\Delta)\Delta(n-1)}{n(a-c+\Delta)^2} > 0 \tag{24}$$ ∎

Moreover, it is straightforward to verify that $\partial\delta^*/\partial n>0$, $\partial\delta^*/\partial a>0$, $\partial\delta^*/\partial c<0$, $\partial\delta^*/\partial\Delta<0$, $\partial\delta^{**}/\partial n>0$, and $\partial(\delta^{**}-\delta^*)/\partial n>0$. This, together with the fact that δ^{**} depends only on n, implies that the comparative statics are the same as under Cournot competition. In addition, the price path under Bertrand competition has a similar shape to that depicted in figure 4. In that case, defection drives prices to marginal costs, hence $p_J^C>p_I^N$.

It is useful to compare the critical discount factors of the Cournot and Bertrand models. We know that in the standard model, collusion is easier to sustain under Cournot competition. Let δ_C^* and δ_C^{**} be the critical discount factors under Cournot competition, and δ_B^* and δ_B^{**} be the critical discount factors under Bertrand competition. Comparing δ_C^{**} and δ_B^{**}, we have that

$$\delta_C^{**}-\delta_B^{**}=\frac{-3n^2+6n+1}{n^3+6n^2+n} \tag{25}$$

which is positive for $n=2$ but negative for $n\geq3$. This means that there is less (more) *TC* under Cournot than under Bertrand competition when $n=2$ ($n\geq3$). Hence, the increase in n from 2 to 3 favors *DD* (relative to *TC*) more under Bertrand than under Cournot competition. This is because the gain from *DD* relative to *TC*, which is given by

$$\Pi^{DD}-\Pi^{TC}=\delta\pi_J^d+\frac{\delta^2}{1-\delta}\pi_J^N-\frac{\delta}{1-\delta}\pi_J^c \tag{26}$$

decreases more under Cournot than under Bertrand competition. In a Cournot market, π_J^d, π_J^N, and π_J^c all decrease with n. Whereas in a Bertrand market, π_J^d and π_J^N are independent of n, hence the decline in $\Pi^{DD}-\Pi^{TC}$ is smaller in a Bertrand market.

Comparing δ_C^* and δ_B^* is more tedious, because of the analytical complexity of δ_C^*. Table 2 gives δ_B^* for the same numerical configurations used in table 1, and allows its comparison with δ_C^*.[10] The table suggests that for many plausible numerical configurations, $\delta_C^*<\delta_B^*$. This means that when $n=2$, $\delta_C^{**}>\delta_B^{**}$ and $\delta_C^*<\delta_B^*$, implying that a Cournot market results in less *TC*, more *DD*, and less *TD* than a Bertrand market.

Table 2. Comparison between the Cournot and Bertrand models

a	n	c	Δ	δ_B^*	δ_B^{**}	$\delta_B^{**}-\delta_B^*$	$\delta_C^{**}-\delta_C^*$
1000	2	50	25	0.475	0.5	0.025	0.146
2000	2	50	25	0.487	0.5	0.013	0.089
1000	5	50	25	0.760	0.8	0.040	0.064
1000	2	55	25	0.475	0.5	0.25	0.146
1000	2	50	20	0.480	0.5	0.020	0.166
1000	20	30	5	0.940	0.95	0.0097	0.0104

When, however, $n\geq3$, $\delta_C^{**}<\delta_B^{**}$ and $\delta_C^*<\delta_B^*$, implying that a Cournot market results in more *TC* and less *TD* than a Bertrand market. The comparison of the extent of *DD* between the two markets is a priori ambiguous. To obtain further insight into the extent of *DD*, table 2 shows $\delta_C^{**}-\delta_C^*$ and $\delta_B^{**}-\delta_B^*$. The length of this interval measures the extent of *DD*. In all the

10 δ_C^* and δ_C^{**} can be read from table 1.

cases considered, $\delta_C^{**} - \delta_C^{*} > \delta_B^{**} - \delta_B^{*}$, suggesting that the extent of *DD* is greater in a Cournot market. In a Bertrand market, *DD* tends to get "squeezed" between *TD*, the incentives for which are high under Bertrand, and *TC*, which obtains for very high values of δ. This reflects the greater incentives for defection in a Bertrand market.

Note that under Bertrand competition, from proposition 6, the extent of *DD* can be written as

$$\delta_B^{**} - \delta_B^{*} = \frac{(\pi_J^d - \pi_J^c) - (\pi_I^d - \pi_I^c)}{\pi_J^d} \tag{27}$$

which represents the gains from defection with the new technology relative to the gains from defection with the old technology. The larger this difference, the greater the incentives for *DD*.

There is less *TD* in a Cournot market, which is consistent with conventional wisdom about collusion under Cournot and Bertrand competition. The new result here is that there is also more *DD* in a Cournot market. This can even result in a situation where there is less *TC* under Cournot than under Bertrand competition.

5. Conclusion

The paper analyzed the relationship between technological collaboration and product market collusion. It was found that technological collaboration facilitates product market collusion, but only in the short-run, by creating an equilibrium where firms collude initially in order to develop the new technology, and then defect afterwards. Nonetheless, there is less defection compared with the standard trigger strategy model. Second, it was shown how the equilibrium depends on the discount factor; in particular, delayed defection is sustained for intermediate ranges of the discount factor. Finally, the comparative statics analysis shows how the equilibrium is affected by changes in the environment. An increase in the scope of the market, an increase in the number of firms, and a lower quality technology all contribute to making collusion more difficult. The explanation for these results lies in the effect of changes in the environment on defection profits and collusive profits. Defection profits increase more than collusive profits in response to an increase in the scope of the market. Similarly, defection profits decline less following an increase in the number of firms than do collusive profits. Whereas permanent collusion is affected only by changes in the number of firms, delayed defection is affected also by the scope of the market and the quality of the new technology.

One major difference between the result obtained here and what is found in the literature is that here technological collaboration can facilitate collusion but only in the short-run: it is a temporary phenomenon. Whereas, in the literature, collusion is sustained permanently. Given the temporary nature of most instances of technological agreements (Kogut, 1989), the finding of endogenous delayed defection is much more appealing.

The model predicts that for intermediate levels of discounting, we should observe some form of market collusion before and during the joint development of technology by firms, and that this collusion should collapse after the collaboration ends. Nonetheless, one should be careful about drawing empirical generalizations from a stylized model like the current one.

Moreover, the result of this model should not be used as an argument against the current policy of encouraging technological collaboration among firms. Rather, it raises the possibility that in some cases technological collaboration, by increasing the stakes, can make collusion easier, albeit only in the short-run.

In a sense, technological collaboration has the same effect as multimarket contact (Scott, 1993; Vonortas, 2000), which makes collusion more valuable and hence more sustainable. One major difference, however, is that the impact of multimarket contact is permanent, whereas the impact of technological collaboration is temporary.

Paradoxically, collusion-enhancing technology cooperation may be beneficial to both firms and society: to firms because it sustains collusion in the short-run and increases future profits (even after collusion has ended), and to society if the gains from the better technology -during but certainly after the collusive period- outweigh the losses from increased market power.

In this model firms face a double prisoner's dilemma: they would gain by both developing the new technology *and* colluding on the output market. Yet, they may end up in a situation where they achieve neither. Technological collaboration can be seen as a partial solution to the prisoner's dilemma. In fact, collusion could be sustained forever (when delayed defection is the equilibrium) if firms agree to develop a new technology between each two periods. This might ensure collusion in each period. Even, firms could go further by subdividing the development of the new technology into several phases, and spreading the development over time.[11] More generally, in situations where delayed defection can arise as an equilibrium, creating positive-sum projects which cover a long time horizon can be a partial solution to the prisoner's dilemma. This ensures that all firms will want to collude for a longer horizon, until the benefits of technological collaboration are exhausted. This can be seen as a justification for strategically delaying the development of new technologies.

In general, changes in demand or technology conditions can be one factor which causes collusion to break down, because it becomes more difficult to agree on the collusive equilibrium, and because it becomes difficult to monitor collusive behavior in a changing environment. The current model proposes another view, that joint technology development can have the opposite effect of that predicted by traditional wisdom. This occurs because the development of the new technology itself is endogenous, and is conditional on collusion.

REFERENCES

Brod, A., and Shivakumar, R. (1997) "R&D Cooperation and the Joint Exploitation of R&D" *Canadian Journal of Economics* **30**, 673-84.

Cabral, L.M.B. (2000) "R&D cooperation and product market competition" *International Journal of Industrial Organization* **18,** 1033-47.

Kultti, K., Takalo, T., and Toikka, J. (2004) *Patents Hinder Collusion.*

Kogut, B. (1989) "The stability of joint ventures: reciprocity and competitive rivalry" *Journal of Industrial Economics* **38**, 183-98.

[11] Nonetheless, this will probably make each instance of collaboration less valuable, and should shift δ^* to the right.

Lambertini, L., Poddar, S., and Sasaki, D. (2002) "Research joint ventures, product differentiation and price collusion" *International Journal of Industrial Organization* **20**, 829-54.

Lin, P. (1998) "Product market competition and R&D rivalry" *Economics Letters* **58**, 105-11.

Martin, S. (1995) "R&D joint ventures and tacit product market collusion" *European Journal of Political Economy* **11**, 733-41.

Rotemberg, J.J., and Saloner, G. (1986) "A Supergame-Theoretic Model of *Price Wars* during *Booms*" *American Economic Review* **76**, 390-407.

Scott, J.T. (1993) *Purposive diversification and economic performance*, Cambridge University Press: N.Y.

Suetens, S. (2003) *Does R&D cooperation facilitate price collusion? An experiment*, Tilburg University, Center for Economic Research, *Discussion Paper* **99**.

van Wegberg, M. (1995) *Can R&D alliances facilitate the formation of a cartel? the example of the European IT industry*, Research Memorandum 004, Maastricht University.

Vonortas, N.S. (2000) "Multimarket contact and inter-firm cooperation in R&D" *Journal of Evolutionary Economics* **10,** 243-71.

INDEX

D

E

F

G

H

I

J

K

L

P

Q

R

S

T